HEAT STROKE

AND

TEMPERATURE REGULATION

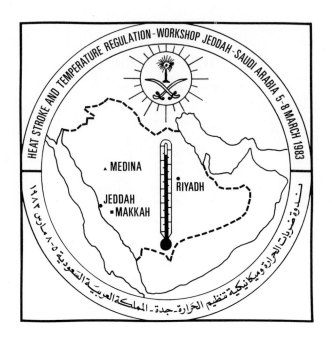

HEAT STROKE

AND

TEMPERATURE REGULATION

Edited by

M. Khogali

Department of Community Medicine
Kuwait University, Kuwait

J. R. S. Hales

Commonwealth Scientific and Industrial Research Organization
Ian Clunies Ross Animal Research Laboratory
Sydney, N.S.W. Australia

ACADEMIC PRESS

A Subsidiary of Harcourt Brace Jovanovich, Publishers
Sydney New York London
Paris San Diego San Francisco São Paulo Tokyo Toronto
1983

ACADEMIC PRESS AUSTRALIA
Centrecourt, 25–27 Paul Street North
North Ryde, N.S.W. 2113

United States Edition published by
ACADEMIC PRESS INC.
111 Fifth Avenue
New York, New York 10003

United Kingdom Edition published by
ACADEMIC PRESS, INC. (LONDON) LTD.
24/28 Oval Road, London NW1 7DX

Printed in Australia

National Library of Australia Cataloguing-in-Publication Data

Heat stroke and temperature regulation.

   Bibliography.
   Includes index.
   ISBN 0 12 406180 X

   1. Heat — Physiological effect.
   2. Heatstroke. 3. Body temperature —
Regulation. I. Khogali, M. (Mustafa).
II. Hales, J. R. S. (Juhl Robert Stanley).

612'.014462

Library of Congress Catalog Card Number: 83-70710

Academic Press Rapid Manuscript Reproduction

Contents

Contributors

Numbers in parentheses indicate the pages on which the authors' contributions begin.

N. M. Abu Al Nasr (109), King Faisal Hospital, Ministry of Health, Taif, Saudi Arabia

Zeinab AbuTaleb (129), King Faisal Hospital, Ministry of Health, Taif, Saudi Arabia

Mohamed S. Al-Adnani (253), Department of Pathology, Faculty of Medicine, Kuwait University, Kuwait

Mohamed Al-D'bbag (171), King Faisal Hospital, Ministry of Health, Taif, Saudi Arabia

Sayed Al Habashi (149), AlShisha Hospital, Makkah, Saudi Arabia

M. I. Al-Khawashki (99), Riyadh Central Hospital, Riyadh, Saudi Arabia

A. Al-Marzoogi (31,293), Forensic Medicine, Ministry of Health, Riyadh, Saudi Arabia

Mustafa Amar (149), AlZahir Hospital, Makkah, Saudi Arabia

Otto Appenzeller (283), Departments of Neurology and Medicine, University of New Mexico, School of Medicine, Albuquerque, New Mexico 87131, USA

Moneim Attia (65, 253), Department of Community Medicine, Faculty of Medicine, University of Kuwait, Kuwait

T. H. Benzinger (53), United States Department of Commerce, National Bureau of Standards, Washington DC 20234, USA

John Bligh (41), Institute of Arctic Biology, University of Alaska, Fairbanks, Alaska 99701, USA

Michel Cabanac (213), Département de Physiologie, Faculté de Médecine, Université Laval, Québec, Province of Quebec, Canada G1K 7P4

K. E. Cooper (79, 189), Department of Medical Physiology, University of Calgary, Calgary, Alberta, Canada T2N 2N4

Adel Nasr El-Din (253), Department of Pharmacology, Faculty of Medicine, Kuwait University, Kuwait

A. El-Ergesus (31), Forensic Medicine, Ministry of Health, Riyadh, Saudi Arabia

Ghalib Elkhatib (253), Department of Community Medicine, Faculty of Medicine, Kuwait University, Kuwait

A. Elkhawad (263), Department of Pharmacology, Faculty of Medicine, Kuwait University, Kuwait

S. F. El-Mahrouky (157), Baljurashy Hospital, Taif, Saudi Arabia

Saad El Sayad (149), Kings Hospital, Medina, Saudi Arabia

H. El-Sayed (99, 129, 149), King Faisal Hospital, Ministry of Health, Taif, Saudi Arabia

Gero Feistkorn (241), Physiologisches Institut Justus-Liebig Universität, D-6300, Giessen, Federal Republic of Germany

ix

Mahmoud Ghallab (171), Directorate of Health Affairs, Makkah, Saudi Arabia
K. Gumaa (109, 119, 157, 253), Department of Biochemistry, Faculty of Medicine, Kuwait University, Kuwait
J. R. S. Hales (223), CSIRO, Ian Clunies Ross Animal Research Laboratory, Prospect, New South Wales 2149, Australia
S. M. Hamdan (263), Department of Pharmacology, Faculty of Medicine, Kuwait University, Kuwait
O. Hassan (263), Department of Pharmacology, Faculty of Medicine, Kuwait University, Kuwait
Claus Jessen (241), Physiologisches Institut Justus-Liebig Universität, D-6300, Giessen, Federal Republic of Germany
M. Khogali (1, 31, 65, 99, 109, 119, 129, 139, 149, 157, 171, 253, 263, 293), Department of Community Medicine, Kuwait University, Kuwait
Alexander R. Lind (179), Department of Physiology, St Louis University, School of Medicine, St Louis, Missouri 63104, USA
Peter Lomax (197), Department of Pharmacology, School of Medicine and the Brain Research Institute, University of California, Los Angeles, California 90024, USA
N. Mahmoud (157), Department of Physiology, Faculty of Medicine, Kuwait University, Kuwait
M. K. Y. Mustafa (99, 109, 119, 157, 253), Department of Physiology, Faculty of Medicine, Kuwait University, Kuwait
Azzam Mutwali (149), King Faisal Hospital, Ministry of Health, Taif, Saudi Arabia
David Robertshaw (13), Department of Physiology and Biophysics, Colorado State University, Fort Collins, Colorado 80523, USA
W. D. Ruwe (79), Department of Medical Physiology, University of Calgary, Calgary, Alberta, Canada T2N 2N4
Salah M. Soliman (129), King Faisal Hospital, Ministry of Health, Taif, Saudi Arabia
O. Thulesius (263), Department of Pharmacology, Faculty of Medicine, Kuwait University, Kuwait
W. L. Veale (79, 189), Department of Medical Physiology, University of Calgary, Calgary, Alberta, Canada T2N 2N4
Hans Gerd Wenzel (273), Institute of Work Physiology, University of Dortmund, D-4600, Dortmund 1, Federal Republic of Germany

Preface

Each year, a great mass of people (currently about two million) from over eighty different nations gathers at Mecca to make the traditional seven-day pilgrimage, the Makkah Hajj. The high radiant heat load and significant aggravation of the thermal environment resulting from the presence of so many bodies in a very restricted area adds to the already high ambient dry bulb temperatures of 35°–50°C. In order to successfully complete the Hajj the pilgrims must carry out rites at certain times. As an example, all two-million pilgrims camp in the eleven square kilometres area of Mina for a three-day period. The Hajj must be one of the greatest human experiments in environmental physiology — the logistics of moving and feeding, and providing sleeping, toilet and medical facilities for, that number of people almost defy comprehension.

Pilgrims at Mount Arafat.

The fact that most of the two million pilgrims successfully performed the Hajj in 1982 is a credit to the Saudi Arabian Government. However, 1190 people were successfully treated for heat illnesses, and several hundred died even before reaching treatment centres. Although this amounts to only about 0.001% of the total number of pilgrims, during an event of this kind only zero is acceptable as the human death rate from preventable circumstances.

Thus, the Saudi Ministry of Health called upon Dr Mustafa Khogali, a man with much clinical and experimental experience in the field, to convene a Workshop on Heat Stroke. Eighteen scientists from different parts of the globe were brought together with approximately 80 clinicians, nurses, technicians and scientists from Saudi Arabia to discuss the problem. This book presents the most recent clinical and experimental observations on heat stroke, together with the most relevant aspects of the basic physiology of temperature regulation and the delegates' recommendations for prevention, maintenance of patients and future research.

We are grateful to the Ministry of Health of Saudi Arabia for sponsoring the Workshop, and to Kuwait University and its Faculty of Medicine for their support. Thanks must also go to the King Fahd Research Centre and the King Abdelaziz University for their assistance and to all the staff of the Health Affairs Directorate, Western Region, Saudi Arabia, whose great efforts made this Workshop a success. Meg Sparshott, Alan Fawcett, Stephen Hales, Moya Frost and Diana Bates are gratefully acknowledged for their assistance in the final preparation of the manuscript.

J. R. S. HALES

Editor's Comments and Recommendations from the Participants

A most vigorous and enlightening discussion persisted throughout the Workshop, with all participants concentrating on the alleviation of heat stroke. Most questions and answers are covered by the papers reported. However, some important points and consequent recommendations follow (with the most relevant chapters indicated).

How Important is Drinking in Hot Environments?

Subnormal hydration of the body is usually also associated with disturbances of the electrolyte balance which can aggravate effects of water deficiency *per se*. Normally, people rarely drink sufficient water to fully replace fluid loss by profuse sweating, and dehydration is likely to be a worse problem with pilgrims as there may be a conscious reduction in fluid intake as a penance, or to reduce the need for toilet facilities, or because of the high cost of purchasing palatable fluids; much water is also used to wash parts of the body five times per day in association with prayers. Hypovolaemia and/or hyperosmolality of the body fluids increases the level of body temperature at which cutaneous vasodilatation occurs and reduces the maximum level reached (Ch. 2 & 20). Sweating rate is also likely to be depressed (Ch. 2). Thus, one's heat tolerance and likelihood of combating heat stress are reduced. However, very recent experimental studies (Ch. 19) reveal that under these conditions, brain temperature may be significantly lower than rectal or oesophageal temperature, probably because sweating on the head may persist at a much higher level than the remainder of the body (with a flow of cooled venous blood from the orbital region to the inside of the skull (Ch. 19). Experiments on sheep (Ch. 22) are planned to further elucidate circulatory, renal and thermoregulatory consequences of dehydration. Dehydration is also aggravated by diarrhoea; this could be due to heat-damage of intestinal flora or to sloughing of mucosal cells, but since it stops of its own accord (requiring no special treatment) it is more likely that an upset to intestinal water and solute absorption is responsible.

What is the Full Significance of Body Temperature Measurement in Patients?

The *prime* diagnostic point of heat stroke is a high body core temperature, probably above 41°C. Thus, the rapid and accurate measurement and subsequent monitoring of this temperature to achieve a fall of 0.3°C per 5 min is of paramount importance. The critical level of temperature for the brain is around 40°-41°C and patients very often have rectal temperatures above this. Since they recover even after rectal temperatures as high as 46°-47°C, questions were raised as to the maximum possible preferential cooling of the brain and whether the rest of the body could be allowed to cook as long as the brain was protected. The maximum recorded difference between hypothalamic (actually tympanic in man) and rectal temperature is 1°C (Ch. 19 & 2), and therefore it is difficult to conceive of a difference of 5°–7°C inferred by the above reports. Nevertheless, measurements of tympanic temperature are now readily available (Ch. 5) and should be used; a safe and simple infra-red device which does away with the slightly risky procedure of touching the tympanic membrane will soon be commercially available. Reliable measurements of skin temperature are also necessary as a guide to the correct adjustment of the cool water sprays and warm air fan of the Makkah Body Cooling Unit to maintain skin temperature around 32°C; in this regard, the monitoring of limb and face skin blood flow would be very valuable. The disproportionately high weighting of 20% given to the facial skin area when deriving mean skin temperature confirms the importance of this body region — but is it due to a critical role in heat dissipation or to 'preferential' neural input?

Is Some Manner of Circulatory Collapse involved in Heat Stroke?

This is not known, however. The fact that many patients are hypotensive, hypovolaemic and unconscious is indicative of circulatory involvement. In animals and human subjects at much lower levels of body temperature (around 42°C and 39.5°C respectively) skin blood flow declines relative to earlier stages of their response to the hyperthermia (Ch. 2, 20 & 25); if this is occurring in our patients, more heat will be conserved and the situation will be aggravated. However, a cutaneous constriction would probably be aimed at conserving central blood volume, venous pressure and cardiac output (Ch. 20 & 25) and therefore what we are dealing with is a competition between cardiovascular and thermoregulatory drives, with the former winning at this stage. Details of circulatory function have been studied during moderate hyperthermia (Ch. 2 & 20), but whether blood vessels in skin of all body regions (particularly the face) constrict like those in limbs is not known, and there have not been comparable studies under such extreme conditions as encountered with heat stroke. The notion of appropriate 'tuning' of the autonomic nervous system was aired (Ch. 25). The picture of vascular collapse and hypoxia in very severe cases, together with signs of consumption coagulopathy may indicate the importance of platelet-released factors such as 5-HT and thromboxane for the pathogenesis and possible therapeutic intervention (use of 5-HT blockers) (Ch. 23).

How Important are Pre-existing Diseases with Respect to Susceptibility to Heat Stroke?

It is quite common for patients with heat stroke to have diabetes, as though they are pre-disposed. At post-mortem, diabetic heat stroke victims do not show petechiael haemorrhages in the pancreas. Since diabetics often exhibit postural hypotension due to faulty baroreflexes, it is conceivable that this could be responsible for relatively poor adjustments in vasculature of various body regions. Also, sweating abnormalities and neuropathy of the autonomic nervous system are common in diabetics. Hypertension or hypotension, hyperthyroidism and probably any peripheral neuropathy could all be quite significant. Although some aspects of effects of thermal status on responses to drugs are well established in animals (Ch. 4), very little is known for humans (Ch. 18). It is certainly easy to imagine marked changes in pharmacokinetics under these conditions, associated both with the high temperature *per se*, and marked changes (probably decrease) in splanchnic and liver blood flow (Ch. 20). It is conceivable that such large temperature changes could alter the characteristics of cardiovascular receptors, for example, changes in the proportion of α-adrenoceptors and β-adrenoceptors.

Are Elderly People More Pre-disposed to Heat Stroke?

Doubtless, a significantly higher proportion of older people are treated. In this population, an age around 50 years is quite old, the majority of people coming from countries where endemic diseases lead to a disproportionately greater physiological than chronological aging. With advancing age (probably starting at 25 years) there is a decline in heat tolerance due to poorer control of peripheral vascular tone and reduced sweating capacity, and there may even be detrimental changes at the level of the CNS (Ch. 17). A man of about 40 years of age has lost about 15% of his maximum aerobic capacity for work compared to when he was 20 years old, and therefore a given job will require a higher proportion of maximum aerobic capacity, which will lead to a greater tendency for a raised body temperature.

Is the Sheep a Suitable Animal Model?

In view of available background data on cardiovascular function in heat-stressed sheep (Ch. 20) and early biochemical and physiological studies of sheep in heat stroke (Ch. 22), the sheep appears to be a suitable model. Even if baboons are readily available (as in some Arabian areas), there would be considerable doubt as to the validity of measurements during severe heat stress because of their very highly emotional state (Ch. 20).

RECOMMENDATIONS

These are made with little consideration of what might be feasible, for example, in view of religious customs.

DEFINE THE PROBLEM

There must be a realistic assessment of the likely numbers of *cases at risk* and *victims*. According to existing statistics and projected meteorological conditions, the present estimate is 1.2–2.4 per one thousand total number of pilgrims, or 2.3 per one thousand foreign pilgrims (who are at greatest risk (Ch. 3).

PREDISPOSING FACTORS

Make every effort possible to alleviate or take into consideration: dehydration; unacclimatization; physical fitness; old age; obesity; chronic disease (endocrine disorders, particularly diabetes, cardiovascular problems, etc.); gastrointestinal, respiratory and fever-producing conditions; use of drugs; and improper clothing.

PREVENTION

1. *In countries of origin.* Provide information concerning climatic conditions in Saudi Arabia. Equip each pilgrim with a certificate (together with the passport) specifying existing illnesses, previous heat illnesses and current therapy. Make sure

each pilgrim has a fan and some sort of head covering, and does not use synthetic underwear.

2. *Instructions en-route (mostly plane or ship).* Using films, oral instruction and leaflets, convey the following information — basic danger of heat exposure , with serious aggravation by dehydration and exercise, and factors such as the added risks in the elderly or in persons with diabetes; most dangerous time of day; minimum daily water intake of four litres (expressed in 'local' units); add plenty of salt to food (but do not take salt tablets).

3. *During the pilgrimage.* Provide numerous shaded rest-places, possibly extending to much of the pilgrimage route, free palatable water at not greater than 200 m intervals, tap water for immediate cooling of heat-exhausted persons (for example, showers), increased number of lavatories (to encourage people to use them more and therefore drink more), and conspicuous flags to the guides for attracting the attention of medical personnel. The previously instructed guides should concentrate on detecting symptoms of impending heat illnesses (which, of course, requires that they be thoroughly educated). Experienced persons should be concentrated along the route at points known to yield a high incidence of heat illnesses.

MANAGEMENT

See papers on management and therapy (Ch. 3, 13 & 15).

Additional to the observance of normal medical practices:

1. *Ambulances.* These should be air-conditioned or at least have a greatly improved high ventilation, staffed by at least a paramedic, immediately take body temperature and then heart rate, if possible make patient drink, and apply external cooling to the patient, if possible by wetting and fanning. Do not apply ice packs to any part of the body (as often done in the past).

2. If no ambulance is available, transport in the semilateral position (rather than, supine or seated); netting stretchers must be provided along the route.

3. *General.* Do not exclude diagnosis if body temperature is below 40°C or if sweating is still apparent. Attach a card to the patient, for recording all treatment. Use Valium only if essential for relaxation (5–10 mg intravenously), and if the patient is having seizures, use barbiturates.

Adopt greater flexibility with respect to continued treatment in the medicare/first aid unit or transfer to Heat Stroke Treatment Centre. A greater number of MBCU trolleys should be provided with each MBCU so the patient can be put directly onto one. Instigate more detailed neurological examination. Raise the legs to improve venous return. Buffer fluid loss with 2 L saline initially plus 1 L if necessary, plus 500 mL low molecular weight dextran; no intravenous glucose unless glucosuria is ruled out by a simple urine test. Display body temperature (preferably tympanic) on a chart recorder (preferably with grid lines at an angle representing a rate of decrease in temperature of 0.3°C per 5 min). Make frequent measurements of all blood gases and continue after cooling . Medical staff must be specially trained.

FUTURE RESEARCH

A. Basic Research

1. There is a need to develop an animal model for examining the changes in bodily systems that lead to death from hyperthermia. Such a study would provide a rational basis for a therapeutic approach aimed at reversing the commonly observed sequelae to hyperthermia. Such a model would be used to examine the following systems and would require the development and involvement of a team of specialists.

(a) Cardiovascular responses to hyperthermia with special reference to the distribution of cardiac output. Particular emphasis should be placed on the vascular beds of the liver, kidney, skin and central nervous system. The relationship of cardiovascular parameters to level of hydration and plasma volume, and to extremes in age should be included. Progress to control of regional vascular responses, although current *in vitro* work (Ch. 23) may continue in parallel.

(b) The metabolic consequences of hyperthermia and the changes in various plasma and CSF constituents that accompany hyperthermia. In addition to the constituents that are routinely measured in any clinical facility, the opportunity should be taken to measure the plasma levels of other substances such as renin, angiotensin, aldosterone, cortisol, triiodothyronine, adrenaline and noradrenaline, bearing in mind the plasma volume changes that occur. In addition, the possible release of endorphins into the circulation and cerebrospinal fluid should be examined.

(c) The respiratory parameters that accompany hyperthermia, particularly a study of diffusion limitations to gaseous exchange; the degree of ventilation/perfusion matching changes in airway resistance, characteristics of the pulmonary circulation.

(d) The changes in renal function that occur during hyperthermia with particular emphasis on total renal blood flow as well as intra-renal distribution.

(e) The changes in function of the central nervous system, using standard neurological techniques. In addition, studies specifically related to metabolism of the brain should be included, for example, the oxygen consumption and substrate utilization of the brain.

(f) The changes in the blood clotting mechanisms that occur during hyperthermia, and the effect of 5-HT blockers on heat stroke.

2. In order to appreciate the most efficient methods of prevention or reversal of hyperthermia, some specific studies on the animal should include:

(a) Studies on the pharmacokinetics of drugs commonly used in medication of the commoner disease conditions, for example, diabetes, hypertension and diseases of the central nervous system, as well as some of the drugs that have been routinely used in the treatment of the hyperthermic patient.

(b) The influence of fever on the course and development of hyperthermia (many heat stroke victims present with a previous history of infection or fever).

(c) The use of preferential head cooling methods to see if the rest of the body can withstand hyperthermia when the temperature of the brain is maintained within certain limits. The effects of too rapid cooling on the integrity of the homeostatic mechanisms. The physiological basis for slow cooling and reversible hyperthermia.

B. Clinical Research

At present the Makkah pilgrimage provides an almost unique opportunity for clinical research into heat stroke and heat disorders in that it is known with some certainty that several hundred cases will be presented at a certain time. In order to maximize the opportunity thus offered, certain clinical studies can be carried out. They would include:

(a) Trials on the effectiveness of certain therapeutic regimens and the drugs commonly used.
(b) Monitoring of bodily processes during the cooling period such as body temperature, blood gases, electrolyte and catecholamine status. Studies on the nature of recurrent hyperthermia.
(c) Monitoring the endocrine changes that accompany hyperthermia and their possible reversal during cooling.
(d) Epidemiological studies on the factors that predispose to heat stroke. The effects on the incidence of heat stroke of the provision of shade, water etc.
(e) Follow-up studies on the subsequent health status of heat stroke patients.
(f) Perform post-mortem examinations to determine the precise cause of death.

1
Heat Stroke: An Overview

M. Khogali

Department of Community Medicine,
Kuwait University, Kuwait*

Exposure to hot environments may result in a variety of heat illnesses - heat cramps, heat exhaustion and heat stroke. Heat stroke is the most serious.

Heat stroke is a symptom complex produced by excessive body heat. It is a medical emergency, but because it presents in different ways, physicians may fail to recognize and treat it promptly. The mortality remains significant, varying from 10-80%.

All ages are at risk - infants, children, the young and the elderly. Infants are at risk if wrapped too warmly during febrile illness or left locked in closed motor cars in hot environments. "Exertion-induced" heat stroke is typically a disorder of the young, healthy athletes, military recruits, or industrial labourers working in hot, humid environments. "Classical" heat stroke is commonly a disorder of the elderly during environmental heat waves. Furthermore drug-induced hyperthermia and heat stroke frequency is on the increase.

Of great importance and uniqueness are the heat stroke casualties encountered during the Holy Annual Makkah Pilgrimage, when it is made during hot months. Hundreds of cases, each presenting differently are seen during a very short period.

This paper is a review of pertinent literature.

*Supported by Research Council Grant No. MC 009, Kuwait University and Kuwait Foundation for the Advancement of Sciences.

HEAT STROKE AND
TEMPERATURE REGULATION
ISBN 0 12 406180 X

In group and crowded situations the exposure to an excessive heat load will often cause several cases of heat illness. According to the Ninth Revision of the International Classification of Disease, there are nine syndromes associated with excess body heat (WHO, 1977). Most important are heat syncope, heat cramps, heat exhaustion and heat stroke; heat stroke is the most serious.

Heat stroke is a symptom complex produced by excessive body heat. It is a medical emergency, but because it presents in different ways, physicians may fail to recognize and treat it promptly (Ralston, 1976). An uncomplicated outcome depends on early diagnosis, proper management and rapid treatment. This treatment in turn depends on an understanding of the pathophysiology of heat stroke.

A wide variety of pre-disposing factors may contribute to the development of heat stroke (Knochel, 1974; Shibolet *et al.*, 1976; Khogali, 1983). In reviews of the literature heat stroke has been associated with chronic illness (Levine, 1969; Shibolet *et al.*, 1976; Collins, 1977), alcoholism (Center for Disease Control, 1981; Kilbourne *et al.*, 1982), pyrexial illness (Wyndham *et al.*, 1965; Austin & Berry, 1956; Minard *et al.*, 1957; Barry & King, 1962), obesity (Minard *et al.*, 1957; Shibolet *et al.*, 1967; Bartley, 1977), fatigue (Bartley, 1977), lack of sleep (Shibolet *et al.*, 1967; Kilbourne *et al.*, 1982), poor physical condition (Minard *et al.*, 1957; Bartley, 1977), and restrictive clothing (Bartley, 1977).

Heat stroke affects both the young and old. The aged and infants seem more susceptible than other groups (Gottschalk & Thomas, 1966; Ellis, 1972; Knochel, 1974; Levine, 1969; King *et al.*, 1981) and men seem twice as likely as women to suffer heat stroke (Bark, 1982*a,b*). "Exertion-induced" heat stroke is typically a disorder of the young healthy athlete or military recruit training in a hot humid environment (Sprung, 1980). "Classical" heat stroke is commonly a dis-order of the elderly occurring during environmental heat waves. These latter persons usually have pre-existing chronic disease or are receiving medication known to predispose to the development of heat stroke.

"CLASSICAL" AND "EXERTION-INDUCED" HEAT STROKE

Hart *et al.* (1982) recently reviewed the clinical charac-teristics of the two forms of heat stroke; Table I, adapted from Hart *et al.*, shows the characteristics.

TABLE I. Characteristics of "Classical" and "Exertion-induced" Heat Stroke[a]

Characteristics	Classical	Exertional
Age	Older	Young
Occurrence	Epidemic form	Isolated cases
Pyrexia	Very high	High
Predisposing illness	Frequent	Rare
Sweating	Often absent	May be present
Acid-base disturbance	Resp. Alkalosis	Lactic acidosis
Rhabdomyolisis	Rare	Common
DIC[†]	Rare	Common
Acute renal failure	Rare	Common
Hyperuricemia	Mild	Marked
Enzymes elevation	Mild	Marked

[a]*Source: Hart et al. (1982).*
[†]*Disseminated intravascular coagulation.*

PHYSIOLOGY

As a background to considering the pathogenesis of heat stroke, it is helpful to consider a few aspects of temperature control. Maintaining normal body temperature is a complicated process involving nearly all organ systems (Johnson & Hales, 1983). Basically it is a balance between the heat accumulation and dissipation mechanisms. The body gains heat either from that produced by metabolism or from the environment if the environmental temperature is greater than body temperature.

Evaporation of sweat is the main method of losing body heat. In man there are about 2.5 x 10^6 sweat glands each innervated by cholinergic pathways. Sweat is formed and secreted from the basal cells of the eccrine sweat glands. Their capacity to secrete water depends on blood flow increasing with cutaneous vasodilatation. The relative ability to sweat per unit body weight decreases with obesity, which decreases the ratio of body surface area to weight (Ralston, 1976). Sweat cools only if it vaporizes. Sweat that is wiped or drips from the body carries an insignificant amount of heat.

Circumstances that inhibit sweating or prevent vaporization may impede heat loss. Anatomic skin abnormalities such as icthyosis, congenital ectodermal dysplasia, scar tissue

from burns, and scleroderma prevent sweating. Occlusive
dressings and clothing do likewise. Decreased cardiac output
from shock, hypotension, dehydration or decreased capillary
flow secondary to direct physical skin cooling or generalized
cutaneous vasoconstriction decreases sweating.

Also, physical constraints that prevent air circulation,
including vapour barrier suits of astronauts, jet pilots and
deep sea divers, and closed non-air conditioned spaces
including auto trunks and semi-trailer trucks have all been
associated with heat stroke (Shibolet *et al.*, 1976). Pharma-
cologic inhibition of sweating is common and occurs with many
drugs (Lomax, 1983).

CLINICAL PATHOLOGIC PICTURE

The classic description of pathological findings in fatal
cases of heat stroke was published by Malamud *et al.* (1946).
It was also reviewed by Knochel (1974) and Shibolet *et al.*
(1976).

Onset

The onset is usually acute (80%) with a few patients having prodromal symptoms and signs lasting minutes to hours
(20%). These include: dizziness, weakness, confusion and
drowsiness, nausea and anorexia, anxiety and headache,
disorientation and disassociation, a staring and apprehens-
ive facial expression (Bark, 1982*a,b*), apathy, irritability
and aggressiveness; irrationality; mania or psychosis; and
tremor, twitching, convulsion, ataxia and cerebellar dysfunction (Bark, 1982*a,b*).

Vital Signs

The body temperature is suddenly elevated to levels not
commonly seen in other conditions. The highest core temperature recorded has been 46.5°C (Khogali *et al.*, 1983) and
47°C (Hart *et al.*, 1982). In our series of heat stroke cases
129 of 172 patients had an initial rectal temperature above
42°C. Only 12 patients had a rectal temperature of 40–41°C
(Al Khawashki *et al.*, 1983). Unfortunately, most clinical
thermometers only measure to 41°C. Special thermometers are
recommended for use in conditions favourable for

precipitation of heat stroke. If the patient is seen some
time after the period of maximal heat production, the temperature may be minimally elevated. This is especially true
when emergency first-aid cooling is carried out. Shibolet *et*
al. (1967) reported heat stroke cases at a very low temperature of 39.5°C. With appropriate treatment the temperature
falls, but occasionally mild temperature elevation occurs
again.

Differential diagnosis of hyperpyrexia should be considered. Pyrexial illness of any kind can contribute to causing
heat stroke, but meningitis, encephalitis and cerebral malaria
may present problems of differential diagnosis (Shibolet *et*
al., 1976; Khogali, 1983).

The skin is usually hot and dry, but in exertion-induced
heat stroke the skin might be wet (Knochel, 1974).

Convulsions and grand mal seizures may occur, otherwise
muscle tone is not increased (Bayley, 1975; Beller & Boyd,
1975; Wyndham, 1977).

Tachypnoea is usually present initially (Al Khawashki *et*
al., 1983). The pulse is usually rapid and weak. Hypotension
and shock are common but not universal.

CNS Disturbances

Disturbances of the central nervous system are present in
all cases. The level of consciousness is often depressed
with coma, stupor or delirium being present. In our series,
deep coma was present in 85% of the patients which were unresponsive to stimuli. Persisting coma in spite of return to
normothermia is a poor prognostic sign. Seizures occur in
approximately 60% of the cases, especially the physically fit.
Signs of cerebellar dysfunction are prominent and may persist. Pupillary changes are frequent, constricted and pin-
pointed pupils were seen in 69% of the patients. Lower motor
neurone and peripheral sensory nerve lesions occur less
frequently. The cerebrospinal fluid is generally crystal
clear and its pressure normal. Complete return of neurologic
functions occurs in most survivors, but cases of persisting
cerebellar dysfunction, hemiplegia, aphasia and emotional
instability are known (Carson, 1972; Mehta & Baker, 1970;
Bark, 1982*a,b*).

Autopsy studies reveal oedema of the brain and meninges
with flattening of the convolutions. Petechiae are common in
the walls of the third and fourth ventricles. Microscopic
examination reveals cortical oedema and congestion with de-
generative changes in the neurones. Cerebellar changes are

more consistent and develop more rapidly than in any other
part of the brain. Suprisingly, the hypothalamus is relatively spared with no light microscopic changes noted (Malamud *et
al.*, 1946; Shibolet *et al.*, 1976).

Cardiovascular

Sinus tachycardia with heart rates of 140 to 150 is the
rule in heat stroke (Kew *et al.*, 1969). Twenty five per cent
of our patients had signs of peripheral circulatory failure
(Al Khawashki *et al.*, 1983). The electrocardiogram may display a flattened or inverted T-wave and various conduction
waves (Kew *et al.*, 1969; Shibolet *et al.*, 1976).

The circulatory collapse seen in heat stroke is not well
understood. The excessive circulatory requirement might be
due to the cutaneous vasodilatation as well as to the reduced
total peripheral resistance as a result of the elevated
metabolic rate and oxygen requirements.

Respiratory

Tachypnoea was the rule in our series of patients. The
hyperventilation, if prolonged, may lead to respiratory
alkalosis and tetany (Shibolet *et al.*, 1976). In 52 out of
223 patients seen during the Makkah Pilgrimage 1982, compensated respiratory alkalosis predominated (30.6%) while
compensated metabolic acidosis (35%) predominated in the
remainder (Gumaa *et al.*, 1983). Levine (1969) reported
clinical signs of pulmonary consolidation in 19 out of 25
elderly patients. Pulmonary oedema is a severe and often
final complication (Al Khawashki *et al.*, 1983).

Renal

In almost all heat stroke cases, there will be protein-
uria with abundant granular casts and red cells in the first
voided urine specimen. In the severe cases acute renal failure is a common complication (Gauss & Meyer, 1917; Malamud
et al., 1946; Vertel & Knochel, 1967; Schrier *et al.*, 1970).
Acute renal failure in heat stroke was first documented in a
review of 125 cases of death from heat stroke during World War
II (Malamud *et al.*, 1946). Several aetiologic factors have
been proposed. Widespread tissue injury suggests direct
thermal injury. Decreased renal plasma flow secondary to

hypotension and peripheral vasodilatation undoubtedly contribute (Ralston, 1976). Secondary disturbances such as
hyperuricemia, acidosis and pigmenturia, probably play a
role. The clinical picture is usually that of acute oliguric
renal failures (Al Khawashki *et al.*, 1983).

Haematologic

Clinical and laboratory evidence of haemorrhagic diathesis is common (Malamud *et al.*, 1946; Shibolet *et al.*, 1967;
Levine, 1969; Kessinger & Rigby, 1970; Stefanini & Spicer,
1971; Mustafa *et al.*, 1983). Petechiae and haemorrhages
occur in all parenchymal organs, in the skin and in the gastrointestinal tract. The haemorrhagic diathesis appears as
purpura, melena, bloody diarrhoea, lung, renal or myocardial
bleeding (Shibolet *et al.*, 1976). Electron microscopic
studies revealed severe capillary endothelial cell damage,
aggregation of platelets and capillary microthrombi in many
organs. Hypoprothrombinaemia, hypofibrinogenaemia, thrombocytopenia and intravascular haemolysis may be found and
suggest dissiminated intravascular coagulation (Wright *et al.*,
1946; Herman & Sullivan, 1959; Knochel *et al.*, 1961;
Mustafa *et al.*, 1983).

Water and Electrolytes

The serum sodium values tend to be low or normal on
admission (Gumaa *et al.*, 1983). In 223 heat stroke patients
seen in the 1982 pilgrimage, the majority were normokalaemic
and hyperglycaemic on admission. Rehydration aggravated the
hyponatraemia, precipitated hypokalaemia but decreased the
hyperglycaemia (Gumaa *et al.*, 1983).

Gastrointestinal

Gastrointestinal symptoms are a common occurrence in heat
stroke (Barry & King, 1962; Carson & Webb, 1973; Shibolet
et al., 1967). In our series of patients diarrhoea was very
common and it was always aggrevated by cooling. Few cases
developed jaundice and these were mainly among the Nigerian
pilgrims.

Musculoskeletal

In cases of heat stroke associated with physical exercise or complicated by generalized seizure activity, necrosis of skeletal muscles and myoglobinuria may occur. It may be accompanied by elevated enzymes of muscle cell origin - CPK, SGOT and LDH.

TREATMENT

Heat stroke is a medical emergency and treatment must be commenced immediately to save the patient's life. The object- ive of the treatment is to reduce the heat load and supplement heat disposal mechanisms. The treatment should include support of vital functions (Khogali *et al.*, 1983) and restor- ation of normothermia using a physiological cooling procedure (Weiner & Khogali, 1980). Measures to prevent convulsions, seizures and shivering should be initiated at once (Khogali & Weiner, 1980). Haemorrhagic diathesis, oliguria and renal failure should be sought and treated. The presence of any pre-existing illness, duration and severity of the thermal insult, and initiation of prompt and efficient, adequate therapy determine the prognosis.

COMMENT

During the last 3 Makkah Pilgrimages 1980-1982, 1567 cases of heat stroke were successfully treated at 6 different Heat Stroke Treatment Centres. The mortality ranged between 5%-18% in the different centres with an overall mortality of 10.5%, which was a remarkable result in the conditions pre- vailing during the Pilgrimage.

Our cases presented in mixed forms of classical and exertion-induced heat stroke. The majority of patients were old but they have to undergo strenuous physical efforts to perform the rites of the Pilgrimage. Pure exertion-induced heat stroke was seen among the young, well built Nigerians who over-exert themselves and walk very long distances under the hot sun. The mortality among them was very high. The meta- bolic status, clinical picture and management procedures followed during the pilgrimage have been reported (Khogali, 1983; Khogali *et al.*, 1983; Gumaa *et al.*, 1983; Al Khawashki *et al.*; 1983; Mustafa *et al.*, 1983).

Acknowledgement

My thanks and gratitude to the Minister of Health, Saudi
Arabia and to all medical personnel for their help and
assistance.

REFERENCES

Al Khawashki, M.I., Mustafa, M.K.Y., Khogali, M. & El-Sayed,
    H. (1983). Clinical presentation of 172 heat stroke
    cases seen at Mina and Arafat. *In* "Heat Stroke and
    Temperature Regulation" (M. Khogali & J.R.S. Hales, eds),
    pp. 99-108. Academic Press, Sydney.
Austin, M.G. & Berry, J.W. (1956). Observations on one
    hundred cases of heatstroke. *J. Am. Med. Assoc.*
    161, 1525-1529.
Bark, N. (1982*a*). The prevention and treatment of heat
    stroke in psychiatric patients. *Hospital and Community*
    Psychiatry 33, 474-476.
Bark, N. (1982*b*). Heat stroke in psychiatric patients. Two
    cases and a review. *J. Clin. Psychiatry 43*, 377-380.
Barry, M.E. & King, B.A. (1962). Heatstroke. *S. African*
    Med. J. 36, 455-461.
Bartley, J.D. (1977). Heat stroke: is total prevention
    possible? *Milit. Med. 142*, 528-535.
Bayley, J.S. (1975). Heat stroke during temperate climatic
    conditions: Case reports. *Milit. Med. 140*, 30-31.
Beller, G.A. & Boyd, A.E. (1975). Heat stroke: A report of
    13 consecutive cases without mortality despite severe
    hyperpyrexia and neurologic dysfunction. *Milit. Med.*
    140, 464-467.
Carson, J. (1972). Heat stroke with left hemiplegia, acute
    tubular necrosis, hypertension and myocardial damage.
    Proc. R. Soc. Med. 65, 752-753.
Carson, J. & Webb, J.F. (1973). Heat illness in England.
    J. Roy. Army Med. Corps 119, 2-83.
Collins, K.F. (1977). Heat illness diagnosis and prevention.
    Practitioner 219, 193-198.
Center for Disease Control (1981). Mortality and Morbidity.
    Weekly Reports *30*, 277-279.
Ellis, F.P. (1972). Mortality from heat illness and heat
    aggravated illness in the United States. *Environ. Res.*
    5, 1-58.

Gauss, H. & Meyer, K.A. (1917). Heat stroke: Report on one
hundred and fifty-eight cases from Cook County Hospital,
Chicago. *Am. J. Med. Sci. 154*, 554-564.

Gottschalk, P.G. & Thomas, J.E. (1966). Heat stroke. *Mayo
Clinic Proc. 41*, 470-482.

Gumaa, K., El Mahsouky, S.F., Mahmoud, N., Mustafa, M.K.Y.
& Khogali, M. (1983). The metabolic status of heat
stroke patients. The Makkah Experience. *In* "Heat Stroke
and Temperature Regulation" (M. Khogali & J.R.S. Hales,
eds), pp. 157-169. Academic Press, Sydney.

Hart, G.R., Anderson, R.J., Crumpler, C.P., Shulkin, A.P.,
Reed, G. & Knochel, J.P. (1982). Epidemic classical heat
stroke: Clinical characteristics and course of 28
patients. *Medicine 61*, 189-197.

Herman, R.H. & Sullivan, B.H. Jr. (1959). Heatstroke and
jaundice. *Am. J. Med. 27*, 154-166.

Johnson, K.J. & Hales, J.R.S. (1983). An introductory
analysis of competition between thermoregulation and other
homeostatic systems. *In* "Thermal Physiology" (J.R.S.
Hales, ed.), in press. Raven Press, New York.

Kessinger, A. & Rigby, P.G. (1970). Hemorrhage and heat
stroke. *Geriatrics 25*, 115-118.

Kew, M.C., Tucker, R.B.K., Bersohn, I. & Seftel, H.C. (1969).
The heart in heatstroke. *Am. Heart J. 77*, 324-335.

Khogali, M. (1983). Epidemiology of heat illnesses during
the Makkah Pilgrimages in Saudi Arabia. *Int. J. Epid.*
(in press).

Khogali, M. & Weiner, J.S. (1980). Heat stroke report on 18
cases. *Lancet ii*, 276-278.

Khogali, M., ElSayed, H., Amar, M., El Sayad, S., Al Habashi,
S. & Mutwali, A. (1983). Management and therapy regimen
during cooling and in the recovery room at different heat
stroke treatment centres. *In* "Heat Stroke and Temperature Regulation" (M. Khogali & J.R.S. Hales, eds),
pp. 149-156. Academic Press, Sydney.

Kilbourne, E.M., Keewhan, C., Jones, S. & Thacker, S.B.
(1982). Risk factors for heat stroke. *J. Am. Med.
Assoc. 247*, 3332-3336.

King, K., Negus, K. & Vance, J.C. (1981). Heat stress in
motor vehicles: A problem in infancy. *Pediatrics
68*, 579-582.

Knochel, J.P. (1974). Environmental heat illness: An
eclectic review. *Arch. Int. Med. 133*, 841-864.

Knochel, J.P., Beisel, W.R., Herndon, E.G., Gerard, L.S. &
Barry, K.G. (1961). The renal, cardiovascular, hemato-
logic and serum electrolyte abnormalities of heat stroke.
Am. J. Med. 30, 299-309.

Levine, J.A. (1969). Heat stroke in the aged. *Am. J. Med.*
47, 251-258.
Lomax, P. (1983). Drug-induced changes in the thermoregulat-
ory system. *In* "Heat Stroke and Temperature Regulation"
(M. Khogali & J.R.S. Hales, eds), pp. 197-211. Academic
Press, Sydney.
Malamud, N., Haymaker, W. & Custer, R.P. (1946). Heat stroke,
a clinico-pathologic study of 125 fatal cases. *Milit.
Surg. 99*, 397-449.
Mehta, A.C. & Baker, R.N. (1970). Persistent neurological
deficits in heat stroke. *Neurology 20*, 336-340.
Minard, D., Belding, H.S. & Kingston, J.R. (1957). Prevention
of heat casualties. *J. Am. Med. Assoc. 165*, 1813-1818.
Mustafa, M.K.Y., Khogali, M., Gumaa, K. & Abu Al Nasr, N.M.
(1983). Disseminated intravascular coagulation among
heat stroke cases. *In* "Heat Stroke and Temperature
Regulation" (M. Khogali & J.R.S. Hales, eds), pp.109-117.
Academic Press, Sydney.
Ralston, R. (1976). Heat Stroke. Minnesota Medicine.
June '76, pp. 411-417.
Schrier, R.W., Hano, J., Keller, H.I., Finkel, R.M.,
Gilliland, P.F., Cirksena, W.J. & Teschan, P.E. (1970).
Renal, metabolic, and circulatory responses to heat and
exercise. *Ann. Intern. Med. 73*, 213-223.
Shibolet, S., Coll, R., Gilet, T. & Sohar, E. (1967). Heat-
stroke: Its clinical picture and mechanism in 36 cases.
Q. J. Med. 36, 525-548.
Shibolet, S., Lancaster, M.C. & Danan, Y. (1976). Heat
stroke: A review. *Aviat. Space & Environ. Med.*
47, 280-301.
Sprung, C.L. (1980). Heat stroke: Modern approach to an
ancient disease. *Chest 77*, 461-462.
Stefanini, M. & Spicer, D.D. (1971). Hemostatic breakdown,
fibrinolysis, and acquired hemolytic anemia in a patient
with fatal heat stroke: Pathogenetic mechanisms. *Am. J.
Clin. Pathol. 55*, 180-186.
Vertel, R.M. & Knochel, J.P. (1967). Acute renal failure due
to heat injury: An analysis of ten cases associated with
a high incidence of myoglobinuria. *Am. J. Med.*
43, 435-451.
Weiner, J.S. & Khogali, M. (1980). A physiological body
cooling unit for treatment of heat stroke. *Lancet i*,
507-509.
World Health Organization (1977). Manual of Internal
Statistical Classification of Diseases, Injuries and
causes of Death. 9th Revision. WHO, Geneva.

Wright, D.O., Reppert, L.B. & Cuttino, J.T. (1946).   Purpuric manifestation of heat stroke.   *Arch. Intern. Med.* *77*, 27-36.

Wyndham, C.H. (1977).   Heat stroke and hyperthermia in marathon runners.   *Annals N.Y. Acad. Sci.* *301*, 128-138.

Wyndham, C.H., Strydom, N.B., Morrison, J.F., Williams, C.G., Bredell, G.A.G., Maritz, J.S. & Munro, A. (1965).   Criteria for physiological limits for work in heat.   *J. Appl. Physiol.* *20*, 37-45.

2
Contributing Factors to Heat Stroke

David Robertshaw

Department of Physiology and Biophysics
Colorado State University
Fort Collins, Colorado

Any factors which either reduce heat loss or increase heat gain will predispose to heat stroke. Exercise may lead to hyperthermia, the degree being a function of work load and not environmental conditions, eg., it will be relatively greater in obese individuals. Exercise hyperthermia is aggravated by conditions that induce haemoconcentration, hypernatraemia and hyperosmotaemia, eg., sweating without fluid replacement. Lack of acclimatization to heat, includes failure to achieve maximal sweating capacity. Therefore, individuals who are not exposed to repetitive heat stress either because they live in temperate regions or spend much time in artificially cooled habitats, will not be acclimatized. Regular physical exercise at a level sufficient to induce sweating, will also induce acclimatization; people whose life style is sedentary are also at greater risk than are acclimatized individuals. Sweating has to be accompanied by an increase in cutaneous blood flow and therefore anything compromising this will impair the effectiveness of cutaneous moisture vaporization. Disorders of the cardiovascular system may affect the redistribution of blood, particularly during exercise where both contracting muscle and skin require an increased blood flow. Many of the conditions noted above can be related to affluence. Superimposed on these circumstances will be some degree of dehydration, whether it be voluntary or involuntary, which will exacerbate the risk of heat stroke.

HEAT STROKE AND
TEMPERATURE REGULATION
ISBN 0 12 406180 X

13

Heat stroke and related disorders, such as heat exhaustion, heat syncope etc., represent a breakdown in the ability of the physiological systems of the body to cope with the heat load. Man has evolved physiological mechanisms to maintain a constant body temperature under a variety of natural (as opposed to man-made) environments. These mechanisms are usually less effective in the newborn, where they are incompletely developed, and in the elderly where they have deteriorated (Robertshaw, 1981; Cooper & Veale, 1983). Thus, at each end of the life span, people are particularly susceptible to accidental hyperthermia. However, in certain circumstances, other factors may predispose to a breakdown in the normal physiological systems and hyperthermia results. The purpose of this review is to examine the physiological adjustments to heat exposure and briefly identify factors that may reduce the efficacy of body temperature control.

COMPONENTS OF THE HEAT BALANCE EQUATION

A. *Magnitude of the heat load*

The heat load on an individual comprises two components, (a) metabolic and (b) environmental.

The metabolic heat load will be elevated by many factors such as prandial state and exercise. Exercise may elevate metabolism by 10- to 20- times that of the resting state. Exercise, therefore, is an important consideration in examining the possible causes of heat disorders.

When air temperature exceeds skin temperature there will be a net flow of heat toward the body which is then added to the metabolic load for dissipation. By far the major heat load experienced under natural conditions, particularly in the tropics and subtropics, is the heat gained by radiation; its physiological significance is considered by Wenzel (1983). Physiologists arbitrarily divide the incident radiation into long-wave and short-wave, the division being based purely on the difference in the optical properties of the skin and clothing. Short-wave radiation comes essentially from the sun and may reach at noon, approximately 800 W/m^2; that is, some 13-times greater than the average resting metabolic rate of 60 W/m^2. In addition to the direct short-wave radiation, there may be short-wave radiation reflected from the surroundings, the amount being a function of the colour or reflectance of the surroundings. For example, in the desert where there is little or no vegetation and where the colour of the soil tends to be lighter, the direct incident radiation may be supplemented by reflected radiation by as much as 400 W/m^2. Long-wave radiation from both the sky and ground may increase the total radiant load to 2200 W/m^2 (Finch, 1972). The absorptance of short-wave radiation is colour dependent, being greater with black than it is with white, but the absorptance of long-wave radiation, particularly by the skin, is independent of colour, i.e. it behaves as a black body (Mitchell, 1974). Two other factors need to be considered when examining the radiant heat load; (a) a black skin will absorb approximately 20% more heat than a white skin (Kerslake, 1972). Although this may seem paradoxical, the epithelial pigment provides protection against the harmful ultraviolet component of solar radiation. (b) although white clothes will absorb less heat than black clothes, it has been demonstrated that the fit of the garment is more important than the colour (Shkolnik *et al.*, 1980), eg., black bedouin garments provide no greater heat load than do white garments, because of the loose fit. However, if the clothing fits tightly against the skin, then colour becomes important. Loose fitting clothing allows adequate convective heat loss, as well as assisting in evaporative heat loss from the skin.

B. *Evaporative heat loss*

As the heat load on an individual increases, the proportion that can be dissipated by convection diminishes and responsibility for heat loss shifts toward evaporation. Since man is not a panting mammal, respiratory evaporative heat loss is minimal and the main mode of evaporation is from the skin. An effective sweating mechanism is, therefore, essential. The heat required for the evaporation of sweat is provided by the blood flowing to the skin, the blood returning from the skin being cooler and helping to maintain the core temperature at a constant level. It is sometimes forgotten that a reduction in cutaneous blood flow will reduce the effectiveness of evaporation and sweat will accumulate on the skin surface. Experiments in which sweating is measured by weight loss (with appropriate corrections for respiratory water loss) may not be equated as measuring sweat secretion. Only in situations in which evaporation is measured and the skin remains dry will the rate of evaporation equal the rate of secretion. Usually this is achieved by ventilated capsule measurements in which the air flow through the capsule is sufficient to maintain a dry skin. Thus, the interpretation of experiments which measure cutaneous blood flow and sweat rate need to be

evaluated in the light of the techniques used. A diminished
cutaneous blood flow during heat exposure will also be
associated with a fall in cutaneous evaporative heat loss and
should not necessarily be interpreted as a reduction in sweat
secretion.

It is obvious that the high heat loads encountered when
an individual exercises under conditions of high air temperature and high radiation require a level of evaporation which
could impose a severe strain on the water content of the body.
Most of the larger mammals are able to rehydrate rapidly
following water loss. Man, however, rehydrates much more
slowly and therefore undergoes "voluntary dehydration", a term
which is defined as the delay in complete rehydration following water loss (Greenleaf & Sargent, 1965). It is known that
this voluntary dehydration results in the complete or partial
breakdown of homeothermy during exercise in the heat. Thus,
Pitts *et al.* (1944) showed that with no water consumption,
rectal temperature rose steadily and the subjects showed heat
exhaustion. When subjects satisfied their thirst, rectal
temperature still rose and exercise performance was still impaired. With forced water ingestion equal to sweat loss, the
increase in rectal temperature was minimal and exercise
performance was excellent. Man, therefore, is particularly
susceptible to high heat loads on account of this relatively
poorly developed mechanism for water homeostasis.

Sweat is hypotonic with respect to blood and the loss of
water is greater than the loss of electrolytes and results in
a rise in the osmolarity of body fluids. The fluid loss
causes a reduction in both plasma volume and interstitial
fluid volume while the intracellular fluid volume remains
essentially intact. Thus, sweating will produce a hyperosmol-
ar hypovolaemia. The relative roles of an increase in osmolarity and a decrease in plasma volume in the cardiovascular
and sweating responses to heat exposure have been the result
of many studies and much speculation.

CARDIOVASCULAR RESPONSES TO A HEAT LOAD

In order to increase both convective and effective
evaporative heat loss from the skin, it is essential that
there be an increase in cutaneous blood flow.

Blood flow to the skin may be increased by a reduction in
sympathetic vasoconstrictor tone or by "active neurogenic
vasodilatation" (see Hales, 1983). In particular, arteriovenous anastomoses are important as they are largely confined

to the skin of extremities and provide large-bore connections
which can bypass the capillary bed to greatly enhance
cutaneous blood flow (Hales, 1981; see Jessen & Feistkorn,
1983). Their functional role in temperature regulation has
been demonstrated by Hales (1981) who has used radioactively-
labelled microspheres injected into the arterial circulation.
Since the microspheres are trapped in the capillary vascular
bed, their appearance on the venous side of the circulation
is an index of an arteriovenous bypass of the capillary bed.
Hales (1981) showed that with heat exposure, there is at
least a 10-fold increase in flow through the anastomoses.
Studies such as these can only be carried out in experimental
animals and although these structures are known to exist in
the skin of man, it can only be speculated that their function
is similar to that of animals.

Changes in limb blood flow of man can be measured using
plethysmography. It has been shown that measurements of total
forearm blood flow during body heating provide an index of
changes in cutaneous blood flow (Roddie *et al.*, 1956). The
major question that has been posed in relation to the control
of cutaneous circulation during heat exposure is, what are the
compensations associated with a reduction in peripheral
vascular resistance caused by an opening up of the cutaneous
vascular beds? For individuals at rest, the pattern seems to
emerge that the increase in cutaneous blood flow is associated
with a rise in cardiac output brought about largely by an increase in heart rate. The increase in heart rate is accompanied by a reduction in blood flow to the splanchnic, renal
and sometimes muscle vascular beds. According to Rowell
(1974), approximately 80 to 85% of the increase in cutaneous
blood flow in humans is brought about by an increase in
cardiac output and 15% is a result of the redistribution of
blood from other regions. If exercise is now superimposed on
the heat stress, other circulatory demands are added, in that
an increase in blood flow to the exercising muscles is
necessary. On the other hand, the increase in heat production
associated with the exercise requires an adequate circulation
to the cutaneous vascular bed for homeothermy; impairment
of cutaneous circulation would limit the rate of heat transfer
from the body core to the skin as well as reducing the effectiveness of cutaneous evaporation which would lead to a rise
in core temperature. Likewise, a reduction in muscle blood
flow at the expense of cutaneous blood flow would result in
anaerobic exercise and severely limit the duration and
intensity of exercise. It is probable that neither vascular
bed is compromised at the expense of another one except under
extreme conditions. Rowell *et al.* (1965) have suggested that

a reduction in splanchnic circulation is sufficient to maintain cutaneous and muscular blood flow during exercise in the heat, however, Bell *et al*. (1983) have very recently found that blood flow in both skin and exercising muscle is compromized in sheep exercising in a mildly warm environment.

Nadel (1980) has reviewed the circulatory adjustments associated with exercise in the heat and identified some of the problems that complicate the process. Cutaneous blood flow requires to be slowed down in order to effect adequate heat exchange. Toward this end, reduced sympathetic venomotor tone leads to filling of the highly compliant capacitance vessels of the skin (Webb-Peploe & Shepherd, 1968). This diversion and pooling of blood in the periphery contributes to the reduction in central blood volume, cardiac filling pressure is reduced and stroke volume declines, requiring a compensatory increase in heart rate in order to maintain cardiac output. In addition, reductions in plasma volume largely brought about by movement of water into exercising muscle which is related to the intensity of the exercise (Nadel *et al*., 1979) result in a further embarrassment of central blood volume. High sweat rates during prolonged exercise will lead to a further decline in the plasma volume.

An increase in cutaneous blood flow will be a function of the thermal demands placed on the body. Nadel *et al*. (1979) have shown that associated with exercise there is a critical threshold core temperature for a sudden increase in cutaneous blood flow. This threshold is reduced, i.e., the increased blood flow occurs at a lower core temperature, if skin temperature is elevated. The high skin blood flow associated with exercise in the heat is attenuated at a critical skin blood flow which suggests that some degree of cutaneous vasoconstriction occurs concurrently with the achievement of maximum heart rate and occurs, presumably, to prevent a further decrease in stroke volume (Fig. 1). It is not known if this heat-induced vasoconstriction is the result of closure of the arteriovenous anastomoses or arteriolar vasoconstriction, or both. However, as a consequence, the rate of transfer of heat from the core to the periphery is reduced and core temperature will rise. An increase in the circulating levels of both epinephrine and norepinephrine during hyperthermia (Robertshaw & Whittow, 1966) may contribute toward the cutaneous vasoconstriction. Thus, under these conditions, circulatory regulation can be said to take precedence over temperature regulation (see also Hales, 1983).

The heat-induced vasoconstriction that accompanies exercise is more readily observed in the upright than in the supine posture, presumably because of the lack of hydrostatic pressure on the peripheral veins (Johnson *et al*., 1974).

Nadel *et al.* (1979) provide an example of one physically unfit subject who failed to show any cutaneous vasconstriction, so that both stroke volume and cardiac output fell, the subject was unable to maintain an adequate blood flow to the muscles and the experiment had to be terminated. Presumably syncope would have resulted if the experiment had been continued. This result stresses the importance of physical fitness as a component of the circulatory response to exercise; unfit subjects are more likely to be susceptible to the possible hypotensive effects of exercise in the heat.

HYDRATION STATE AND THERMOREGULATORY RESPONSES TO A HEAT LOAD

A. *Circulatory adjustments*

Since a reduction in plasma volume appears to be an important part of the response to exercise in the heat, it might be anticipated that the level of hydration would have an impact. The studies of Horstman & Horvath (1972) clearly showed that there was a reduction in forearm blood flow associated with dehydration in resting subjects exposed to heat and some degree of hyperthermia resulted. When forearm blood flow is related to core temperature (oesophageal temperature) it can be seen that the increase in forearm blood flow, which is a linear function of oesophageal temperature, is affected by dehydration in that there is an increase in the threshold temperature at which forearm blood flow increases, but there is no change in the slope of that relationship (sensitivity) (Fig. 1). Furthermore, in the dehydrated state, there is evidence of relative cutaneous vasoconstriction at levels of blood flow which are much lower than those in the hydrated state. This suggests that dehydration produces a central suppression of the vasodilator outflow to the cutaneous blood vessels. Furthermore, the results shown in Fig. 1 suggest that the reduction in plasma volume associated with dehydration may induce a marked reduction in cutaneous blood flow as opposed to an attenuation of the blood flow-temperature relationship.

Whether or not the effect of dehydration is associated with changes in volume, or is a consequence of an increase in osmolarity is also shown in Fig. 1. Under conditions in which either the osmolarity or plasma volume are changed, it can be seen that both stimuli are additive. Presumably, therefore, both osmoreceptors and volume receptors may have a

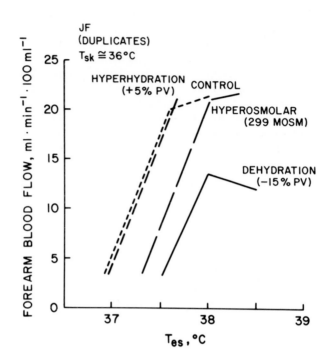

Fig. 1.   Forearm blood flow as a function of oesophageal temperature under differing conditions.   From Nadel (1980), courtesy of the Federation of American Societies for Experimental Biology.

role in control of the peripheral circulation.   The experiments with hyperhydration in which plasma volume is increased (usually by water ingestion and antidiuretic hormone administration) suggests that peripheral vasoconstriction is a function of the degree of hypovolaemia resulting from the exercise.

B.   Sweating

Activation of sweat glands in humans is the result of stimulation of sympathetic fibres to the sweat glands.   The magnitude of the response to neural stimulation is a direct function of skin temperature (Nadel *et al.*, 1971).   Thus, in studies involving the control of sweating, it is important that both core and skin temperature be known.   For example,

during exercise the skin temperature over exercising muscles
may be higher than that in other regions of the body and the
sweating rate of these regions may be enhanced. This is
usually manifest as an increase in the slope of the relationship between core temperature and sweat rate (Wells &
Buskirk, 1971). During dehydration there appears to be a
reduction in overall sweat rate with a resultant hyperthermia
(Ekblom *et al.*, 1970). It has been suggested that the reduction in sweating rate following a decrease in blood volume is
due to the increase in osmotic pressure of the plasma. Thus,
Senay (1979) has shown that there are negative relationships
between sweat rate and both plasma osmolarity and plasma
sodium concentration. Since most of the osmolarity is due to
the concentration of sodium ions, this relationship is not un-
expected. Experiments on animals have shown that changes in
tonicity without changes in plasma sodium concentration are
sufficient to suppress evaporative heat loss and raise body
temperature during exercise in the heat (Kozlowski *et al.*,
1980). Likewise, Doris & Baker (1982) have shown that the
administration of water into the lateral cerebral ventricles
of the dehydrated cat causes an immediate restoration of
evaporative heat loss and fall in body temperature. All these
studies point to a specific connection between central osmoreceptors and the efferent outflow from the hypothalamus.
Experiments on both man and animals clearly indicate that the
effects of dehydration on evaporative heat loss only apply
under the conditions of ambient or exercise heat load, i.e.,
when there is a central drive to the evaporative heat loss
mechanisms; body temperature under thermoneutral conditions
is normal. The role of hypovolaemia without changes in osmolality, that is, an isotonic contraction of the vascular
space, has been studied by Fortney *et al.* (1981). They have
shown the slope of the relationship between sweat rate and
core temperature is significantly reduced when measurements
are made from the chest and arm. However, there was no change
when measurements were made over actively contracting muscles;
therefore, presumably the high local skin temperatures offset
the central inhibition in a manner already referred to. Their
studies provide evidence that volume receptors, such as the
stretch receptors in the right atrium, may have neural projec-
tions to the temperature regulating centres of the hypothalamus. It is generally conceded that antidiuretic hormone,
which would be released during an isotonic reduction of blood
volume, has no effect on sweat rate (Pearcy *et al.*, 1956;
Senay & Van Beaumont, 1969).

   Thus, dehydration whether it results in either hypovolaemia or an increase in osmolarity or a combination of both,

produces a different response in the vascular system than it does in the evaporative heat loss system: In the former, there is good evidence for a shift in the threshold of core temperature for the onset of vasodilatation and no change in the gain of the relationship, whereas in the evaporative heat loss system there is simply a depression in the gain with no change in the threshold. It may be that expressing sweat rate purely in terms of core temperature is an over-simplification and other temperatures may have a greater modifying influence on the sweat rate-core temperature relationship than they do on the blood flow-core temperature relationship. If in fact there is an elevated threshold for the onset of sweating, this might not be apparent if only one estimate of core temperature is made. It is known, for example, that in mammals which pant, dehydration results in an elevation of body temperature but brain temperature is to some extent preserved. This is achieved by the use of a counter-current heat exchanger in the brain, the *rete mirabile*. The arterial blood supply to the brain is cooled by cool blood draining from the nasal area, the nasal mucosa itself being cooled by panting. No such system exists in primates comparable to the *rete mirabile* of panting animals (Baker, 1979). However, even in sub-human primates where the base of the brain is supplied by the internal carotid artery and not a *rete mirabile*, cooling the skin of the head would also cool the base of the brain (Baker & Fuller, 1982). There is a possibility, therefore, that such a system may exist in man.* Core temperature measurements, whether they be measurements of rectal temperature or oesophageal temperature may, under conditions of dehydration, give a misleading impression of thermal inputs to the hypothalamus.*

ACCLIMATION AND PHYSICAL FITNESS

Repeated exposure to heat stress over a period of 10 to 14 days produces physiological changes which improve the ability of individuals to tolerate heat. Reversion to the unacclimated state occurs within a few weeks (Williams *et al.*, 1967). Part of the acclimation process involves an increase in the maximum sweat rate (Table I). Improvement in sweating

*Cabanac, in this Workshop, discusses such preferential brain cooling in man, including influences of dehydration and exercise (Editors' Note).

TABLE I.   *Maximum sweat rates (g m^{-2}h^{-1}) of acclimated and unacclimated men and women exercising in the heat (Wyndham et al., 1965)*

	Women	Men
Unacclimated	232	334
Acclimated	385	602
Ratio	1.66	1.80

[handwritten annotations: beside Women "}65%"; "80% {"; beside Men "334 >40%", "602 >56%"; below "prop. increase"]

appears to be almost entirely due to an increase in the ability of the individual glands to produce sweat. Thus, Collins *et al.* (1965) have been able to locally acclimate sweat glands by the repeated stimulation of sweating by local injection of sudorific drugs. Localized acclimation can also be inhibited by prolonged use of anhidrotic drugs (Goldsmith *et al.*, 1967). Since in the acclimated individual, stimulation of sweating occurs at lower skin or core temperatures than in non-acclimated subjects (Wyndham, 1967), the general conclusion is that sweat gland acclimation involves not only an enhanced hypothalamic outflow but also a greater sweat gland responsiveness to neural stimulation.

Table I also demonstrates the well known difference between males and females in their intrinsic sweating capabilities. This difference appears to reside in a lower sweat gland density in females than in males (Foster *et al.*, 1969).

PHYSICAL CONDITIONING

Improvement of physical condition appears to improve the ability of individuals to tolerate heat exposure. This may be, in part, due to some degree of heat acclimation and in part to improved responses of the cardiovascular system to heat exposure. Since exercise imposes a heat load on the body which must be dissipated in order to prevent hyperthermia, it has been suggested that physical conditioning will result in repeated stimulation of heat loss mechanisms and thereby produce heat acclimation (Piwonka *et al.*, 1965; Piwonka & Robinson, 1967). However, it has subsequently been shown that this is only partially true; acclimation resulting from exercise is beneficial only in cases of short-term heat exposure, there being no difference during long-term heat exposure (Strydom & Williams, 1969). Senay (1975) has pioneered another approach to this problem, by investigating the changes in plasma volume that accompany both heat acclimation and physical conditioning. Heat exposure of resting individuals has been shown to produce haemodilution. This is due in part to a fall in capillary hydrostatic pressure due to cutaneous vasodilatation, and in part to the movement of proteins into the vascular compartment. This expansion of the plasma volume allows a decline in plasma volume to occur during heat exposure before the need to resort to compensatory cardiovascular adjustments. Heat acclimation results in two alterations in fluid shift that help to stabilize the cardiovascular system during heat exposure. Firstly, the rate of protein transfer and rate of water movement into the circulation is accomplished more rapidly, and secondly, the resting plasma volume is expanded, the increase being largely due to the movement of protein from the interstitial spaces via the lymphatic vessels (Senay & Kok, 1977).

Physical conditioning has been shown to increase plasma volume by up to 22% (Brotherhood *et al.*, 1975; Williams *et al.*, 1979) which would therefore confer a beneficial effect on the response to heat exposure. However, the lack of any sustained improvement to heat stress appears to be due to incomplete acclimation of the sweat glands (Senay & Kok, 1977). Presumably, the inability to achieve high skin temperatures by exercise under cool conditions prevents the development of the local component in sweat gland acclimation. The expansion of plasma volume by exercise may be due, in part, to an increase in antidiuretic hormone and angiotensin as well as an increase in the plasma albumin content (Convertino *et al.*, 1980).

SUMMARY

This review of some of the physiological changes that accompany heat exposure allows an identification of some of the factors that will predispose to a breakdown in the ability to lose heat. They may be listed as follows:

(1) Dehydration, which under conditions of high fluid loss may be involuntary due to the relatively poor fluid homeostatic mechanisms of man, i.e., a low thirst drive. This will embarass the cutaneous circulation as well as the sweat secretion systems.

(2)   Lack of heat acclimation is accompanied by both poor regulation of fluid volume shifts as well as an incompletely developed sweating capability.

(3)   Poor physical fitness not only makes exercise less efficient but also fails to confer an adequate circulatory base for the cardiovascular adjustments to heat.

Individuals at risk therefore are those who either voluntarily undergo dehydration or who are unable to replenish their thirst. Likewise, any disorders, such as diarrhoea or vomiting that result in dehydration, will exacerbate their predicament under hot conditions. In addition, cardiovascular disorders and conditions such as diabetes mellitus which tend to predispose to cardiovascular disease will reduce the ability to lose heat. Individuals who are from temperate countries will be unlikely to be heat acclimated and a sedentary life-style will preclude the development of fluid volume adjustments that accompany physical conditioning and heat acclimation. Sedentary, obese individuals will have difficulty not only in sustaining any level of physical exercise but their thermoregulatory responses will be inadequate. If old age is superimposed on this list of factors, it can be seen that careful monitoring of high risk individuals will be required to offset a breakdown in effective heat loss mechanisms.

REFERENCES

Baker, M.A. (1979). A brain-cooling system in mammals. *Scientific American 240*, 114-122.

Baker, M.A. & Fuller, C.A. (1982). Differential regulation of brain and body temperature in the squirrel monkey. *Fed. Proc. 41*, 1695.

Bell, A.W., Hales, J.R.S., King, R.B. & Fawcett, A.A. (1983). Influence of heat stress on exercise-induced changes in regional blood flow in sheep. *J. Appl. Physiol.* (in press).

Brotherhood, J., Brozovic, B. & Pugh, L.G.C. (1975). Haematological status of middle- and long-distance runners. *Clin. Sci. Mol. Med. 48*, 139-145.

Collins, K.J., Crockford, G.W. & Weiner, J.S. (1965). Sweat gland training by drugs and thermal stress. *Arch. Envir. Hlth.11*, 407-422.

Convertino, V.A., Greenleaf, J.E. & Bernauer, E.M. (1980). Role of thermal and exercise factors in the mechanism of hypervolemia. *J. Appl. Physiol. 48*, 657-664.

Cooper, K.E. & Veale, W.L. (1983). The elderly and their risk of heat illness. *In* "Heat Stroke and Temperature Regulation" (M. Khogali & J.R.S. Hales, eds), pp.189-196. Academic Press, Sydney.

Doris, P.A. & Baker, M.A. (1982). Intracranial osmoreceptors control evaporation in the heat-stressed cat. *Brain Res. 239*, 644-648.

Ekblom, B.C.J., Greenleaf, C.G., Greenleaf, J.E. & Hermansen, L. (1970). Temperature regulation during exercise dehydration in man. *Acta Physiol. Scand. 79*, 475-483.

Finch, V.A. (1972). Energy exchanges with the environment of two East African antelopes, the eland and the hartebeest. *Symp. Zool. Soc. Lond. 31*, 315.

Fortney, S.M., Nadel, E.R., Wenger, C.B. & Bove, J.R. (1981). Effect of blood volume on sweating rate and body fluids in exercising humans. *J. Appl. Physiol. 51*, 1594-1600.

Foster, K.G., Hey, E.N. & Katz, G. (1969). The response of the sweat glands of the new-born baby to thermal stimuli and to intradermal acetylcholine. *J. Physiol. (Lond.) 203*, 13-29.

Goldsmith, R., Fox, R.H. & Hampton, I.F.G. (1967). Effects of drugs on heat acclimatization by controlled hyperthermia. *J. Appl. Physiol. 22*, 301-304.

Greenleaf, J.E. & Sargent, F. (1965). Voluntary dehydration in man. *J. Appl. Physiol. 20*, 719-724.

Hales, J.R.S. (1981). Thermoregulatory implications for partition of the circulation between nutrient and non-nutrient circuits. *In* "Progress in Microcirculation Research" (D. Garlick, ed.), pp. 316-328. Univ. of NSW Committee in Postgrad. Med. Edn, Sydney.

Hales, J.R.S. (1983). Circulatory consequences of hyperthermia: An animal model for studies of heat stroke. *In* "Heat Stroke and Temperature Regulation" (M. Khogali & J.R.S. Hales, eds), pp.223-240. Academic Press, Sydney.

Horstman, D.H. & Horvath, S.M. (1972). Cardiovascular and temperature regulatory changes during progressive dehydration and euhydration. *J. Appl. Physiol. 33*, 446-450.

Jessen, C. & Feistkorn, G. (1983). Some aspects of cutaneous blood flow and acid-base balance during hyperthermia. *In* "Heat Stroke and Temperature Regulation" (M. Khogali & J.R.S. Hales, eds), pp.241-252. Academic Press, Sydney.

Johnson, J.M., Rowell, L.B. & Brengelmann, G.L. (1974).
Modification of the skin blood flow-body temperature
relationship by upright exercise. *J. Appl. Physiol.*
37, 880-886.
Kerslake, D.McK. (1972). "The Stress of Hot Environments".
Cambridge University Press.
Kozlowski, S., Greenleaf, J.E., Turlejska, E. & Nazar, K.
(1980). Extracellular hyperosmolality and body temper-
ature during physical exercise in dogs. *Am. J. Physiol.*
239, R180-183.
Mitchell, D. (1974). Physical basis of thermoregulation. *In*
"MTP International Review of Science", Physiology Series
1, Vol. 7 (D. Robertshaw, ed.), pp. 1-32. Butterworths,
London.
Nadel, E.R. (1980). Circulatory and thermal regulations
during exercise. *Fed. Proc. 39*, 1491-1497.
Nadel, E.R., Bullard, R.W. & Stolwijk, J.A.J. (1971).
Importance of skin temperature in the regulation of
sweating. *J. Appl. Physiol. 31*, 80-87.
Nadel, E.R., Cafarelli, E., Roberts, M.F. & Wenger, C.B.
(1979). Circulatory regulation during exercise in
different ambient temperatures. *J. Appl. Physiol.*
46, 430-437.
Pearcy, M., Robinson, S., Miller, D.I., Thomas, J.T. &
De Brota, J. (1956). Effects of dehydration, salt
depletion and pitressin on sweat rate and urine flow.
J. Appl. Physiol. 8, 621-626.
Pitts, G.C., Johnson, R.E. & Consolazio, F.C. (1944). Work
in heat as affected by intake of water, salt and glucose.
Am. J. Physiol. 142, 253-259.
Piwonka, R.W., Robinson, S., Gay, V.L. & Manalis, R.S. (1965).
Preacclimatization of men to heat by training. *J. Appl.
Physiol. 20*, 379-384.
Piwonka, R.W. & Robinson, S. (1967). Acclimatization of
highly trained men to work in severe heat. *J. Appl.
Physiol. 22*, 9-12.
Robertshaw, D. (1981). Man in extreme environments, problems
of the newborn and elderly. *In* "Bioengineering, Thermal
Physiology and Comfort" (K. Cena & J.A. Clark, eds),
pp. 169-179. Elsevier, Amsterdam.
Robertshaw, D. & Whittow, G.C. (1966). The effect of hyper-
thermia and localized heating of the anterior hypothalamus on the sympatho-adrenal system of the ox (*Bos
taurus*). *J. Physiol. (Lond.) 187*, 351-360.

Roddie, I.C., Shepherd, J.T. & Whelan, R.F. (1956). Evidence from venous oxygen saturation measurements that the increase in forearm blood flow during body heating is confined to the skin. *J. Physiol. (Lond.) 134*, 444–450.

Rowell, L.B. (1974). Human cardiovascular adjustments to exercise and thermal stress. *Physiol. Rev. 54*, 75–159.

Rowell, L.B., Blackmon, J.R., Martin, R.H., Mazzarella, J.A. & Bruce, R.A. (1965). Hepatic clearance of indocyanine green in man under thermal and exercise stresses. *J. Appl. Physiol. 20*, 384–394.

Senay, L.C. (1975). Plasma volumes and constituents of heat-exposed men before and after acclimatization. *J. Appl. Physiol. 38*, 570–575.

Senay, L.C. (1979). Temperature regulation and hypohydration: A singular view. *J. Appl. Physiol. 47*, 1–7.

Senay, L.C. & Kok, R. (1977). Effects of training and heat acclimatization on blood plasma contents of exercising men. *J. Appl. Physiol. 43*, 591–599.

Senay, L.C. & Van Beaumont, W. (1969). Antidiuretic hormone and evaporative weight loss during heat stress. *Pflügers Arch. 312*, 82–90.

Shkolnik, A., Taylor, C.R., Finch, V.A. & Borut, A. (1980). Why do Bedouins wear black robes in hot deserts? *Nature 283*, 373–375.

Strydom, N.B. & Williams, G.C. (1969). Effect of physical conditioning on state of heat acclimatization of Bantu laborers. *J. Appl. Physiol. 27*, 262–265.

Webb-Peploe, M.M. & Shepherd, J.T. (1968). Response of dog's cutaneous veins to local and central temperature changes. *Circ. Res. 23*, 693–699.

Wells, C.L. & Buskirk, E.R. (1971). Limb sweating rates overlying active and nonactive muscle tissue. *J. Appl. Physiol. 31*, 858–863.

Wenzel, H.G. (1983). Contribution of air humidity and heat radiation to heat stress due to elevated air temperature. *In* "Heat Stroke and Temperature Regulation" (M. Khogali & J.R.S. Hales, eds), pp.273–282. Academic Press, Sydney.

Williams, C.G., Wyndham, C.H. & Morrison, J.F. (1967). Rate of loss of acclimatization in summer and winter. *J. Appl. Physiol. 22*, 21–26.

Williams, E.S., Ward, M.P., Milledge, J.S., Witley, W.R., Older, M.W.J. & Forsling, M.L. (1979). Effect of the exercise of seven consectuve days hill-walking on fluid homeostasis. *Clin. Sci. 56*, 305–316.

Wyndham, C.H. (1967). Effect of acclimatization on the sweat rate/rectal temperature relationship. *J. Appl. Physiol. 22*, 27–30.

Wyndham, C.H., Morrison, J.F. & Williams, C.G. (1965).   Heat
    reactions of male and female caucasians. *J. Appl.
    Physiol. 20*, 357–364.

3
Organizational Set Up: Detection, Screening, Treatment and Follow-up of Heat Disorders

A. Al-Marzoogi,[1] M. Khogali[2] and A. El-Ergesus[1]

Ministry of Health, Saudi Arabia[1]

Department of Community Medicine
Faculty of Medicine
Kuwait University, Kuwait[2]

Approximately two million muslim pilgrims are at risk from heat disorders when they perform the Hajj in hot weather. In 1982, 1119 heat stroke and 4000 heat exhaustion cases were managed at special centres in the Saudi Ministry of Health institutions. More cases are expected to occur in the future.

The programme designed by the Saudi Health Authorities takes into consideration the medical facilities available along the pilgrimage route, specific characteristics of the divergent multi-national pilgrims and the multiple predisposing factors.

Raising awareness by doctors, nurses and allied personnel is imperative. The role played by Islamic Medical Mission and their co-operation is emphasized as well as the role of other agencies.

Special forms for early detection and screening of cases of hyperpyrexia are used at outer posts and receptions of hospitals. Specific first aid and life-saving measures are followed prior to reaching treatment centres, especially during transport of patients.

Essentially 100% of heat exhaustion cases recovered and a significant low mortality was achieved with heat stroke. Late complications and results of follow-up of patients are

*unclear because of the mobility of the pilgrims. Morbidity due to heat-induced illness has not been properly documented.
    Despite the enormous problems of such over-crowding, the programme attained significant success; shortcomings are rectifiable.*

Over the last two decades there has been a drastic increase in the number of pilgrims to Makkah (Fig. 1) –
approximately two million foreign and resident pilgrims performed the Hajj last year. In 1951 only one hundred thousand

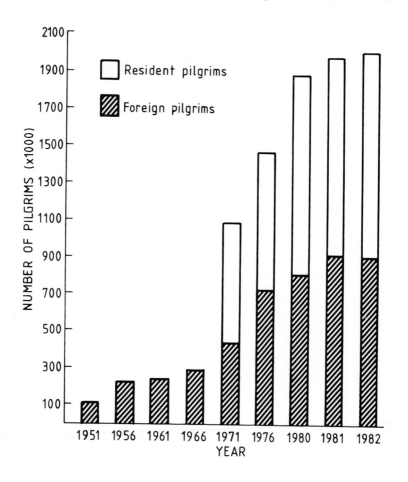

Fig. 1.   Number of foreign and resident pilgrims, 1951-82.

foreign pilgrims performed the pilgrimage compared with 294 thousand in 1966 and 900 thousand in 1981 - a nine-fold increase. Their mode of travel changed significantly (Fig. 2) In 1951 only 14% travelled by air, 2% by land and 84% by sea; in 1981 the proportions were 60%, 32% and 8% respectively.

The change in mode of travel shortened the incubation period, and lessened the acclimatization of pilgrims. When this is added to the problems of over-crowding and sanitation, an unsurmountable health problem is created. When the pilgrimage occurs in the hot weather the problem is aggravated. Heat disorders and heat-induced illnesses become the major problem.

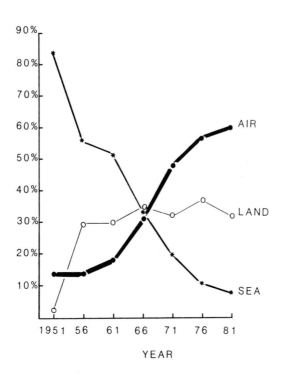

Fig. 2. Mode of travel of foreign pilgrims, 1951-81.

Size of the Problem

Between 1982 and 2000 (1402 AH – 1420 AH) the Annual Makkah Pilgrimage will occur during the hot cycle September to May. A very vast and wide range of heat disorders: (i) heat syncopy; (ii) heat cramps; (iii) heat exhaustion due to water depletion; (iv) heat exhaustion due to salt depletion; (v) heat exhaustion, unspecified; (vi) heat stroke; are expected to be seen. Heat-induced illnesses and chronic diseases aggravated by the high environmental temperatures would be very common.

The number of cases of heat stroke admitted and treated along the pilgrimage route in 1982 and the number of expected cases of heat exhaustion are shown in Table I.

For each case of heat stroke, 5–8 cases of heat exhaustion are expected (Ellis, 1976).

Since management of heat exhaustion was performed at the field station, at health centres distributed along the pilgrimage route and at medical centres run by the Islamic Medical Missions, it was very difficult to give an accurate number of the heat exhaustion cases seen.

Also, a number of patients die before they reach the hospital and in the prevailing hot environmental conditions our contention, since postmortem examinations are not allowed, is that they died of heat stroke or heat-aggravated illnesses.

Medical Facilities

The Ministry of Health of Saudi Arabia provide free medical care to all pilgrims during their stay in the Kingdom.

TABLE I. Heat Stroke Cases Seen in 1982 and the Expected Cases of Heat Exhaustion

Place	Heat stroke cases admitted	Heat exhaustion expected cases	
		Min.	Max.
Makkah	709	3545	5672
Mina	169	845	1352
Arafat	53	265	424
Medina	188	940	1504
	1119	5595	8952

Health services are provided at points of entry, whether by air, land or sea and along the pilgrimage route. Full mobilization of all the medical personnel is implemented during the pilgrimage season. Field stations are distributed among the tents in Arafat for easy access. Health Centres are located along the pedestrian routes between Arafat and Mina and strategically placed in Mina. Hospitals and polyclinics are within easy access along the motor road joining Makkah to Medina.

Plan of Management of Heat Disorders

In view of the size and seriousness of the problem the Ministry of Health initiated plans and programmes aiming at early detection, treatment and prevention of heat illness. The authors have been involved from the start in discussion and formulation of the plan.

The plan takes into consideration the medical facilities available during the pilgrimage. It is based on raising the awareness of doctors, nurses and allied personnel to the problem. Special training programmes are conducted for all medical personnel involved in work during the pilgrimage. Doctors working in all medical units receiving patients, undergo a short training course to acquaint them with the diverse aspects of the problem and the procedures to be followed in general and specifically in managing cases of heat stroke and heat exhaustion (Khogali, 1982).

A special heat illness form (Fig. 3) was designed and circulated to all doctors in the field stations, health centres and receptions in the main hospitals. Its objective when properly completed is to help the doctors in screening and early detection of the cases. In our experience the Heat Illness form is very useful.

Fig. 4 shows a flow chart of the plan of management as was operated during the 1982 pilgrimage. All cases of hyperpyrexia are received at the reception at hospitals or at the health centres. Trained personnel would screen all hyperpyrexial cases and complete form A (Fig. 3).

Management at Hospitals

Each hospital along the pilgrimage route has a Heat Stroke Treatment Centre (HSTC) with varying numbers of Makkah Body Cooling Units (MBCU). In each hospital there is a special ward for treatment of heat exhaustion cases

KINGDOM OF SAUDI ARABIA
MINISTRY OF HEALTH

DIRECTORATE OF HEALTH
WESTERN REGION. MAKKAH.

HEAT ILLNESS FORM

SCREENING, DETECTION & EARLY MANAGEMENT

A. PERSONAL DATA

B. GENERAL DATA

1. NAME _____
2. AGE _____
3. SEX _____
4. NATIONALITY_____
5. OCCUPATION _____

6. DATE _____
7. PLACE _____
8. REF. PHYSN. _____
9. TIME OF EXAM _____
10. MUTAWIF NAME _____

PRESENTING SYMPTOMS OF SIGNS (TICK   Where appropriate) .

C. PRESENTATION (GENERAL)

1. CONSCIOUSNESS   DURATION (1   HRS)
   ☐ CONSCIOUS _____
   ☐ SEMICOMA _____
   ☐ COMA _____
   ☐ CONVULSIONS_____

2. CIRCULATION
   ☐ NORMAL   ☐ SHOCK
3. RESPIRATORY
   ☐ NORMAL   ☐ STRESS
4. SKIN   ☐ FLUSH   ☐ PETECHAE
5. GASTRO—INTESTINAL :
   ☐ BLEEDING   ☐ DIARRHOEA
            ☐ VOMITING

CLINICAL EXAMINATION :

1. PULSE _____ per minute
2. B. P. _____ SYS _____ DIAS_____
3. TEMP oC : REC : AX _____ ORAL ____
4. PUPILS : ☐ Constr. **REACTION** ☐ + ve
            ☐ Dilated   ☐ – ve
6. SKIN☐ RY☐ WET ☐ HOT ☐ COLD
7. NECK REGIDITY ☐ PRESENT ☐ ABSENT
8. SIGNS OF LATE REALIZATION : ☐
9. CHEST :   ☐ CLEAR
10. ABDOMEN :☐ NORMAL ☐ ABNORMAL
11. SKELETAL : MUSCLE REGIDITY ☐

E. OTHER OBSERVATIONS _____
F. MANAGEMENT :   1) _____
                  2) _____
G. REFERENCE ☐ YES ☐ NO   TO _____

NAME & SIGNATURE :

Fig. 3.   Heat Illness Screening Form.

(Al Dabbag *et al.*, 1983). On screening the patient, if it is
found that he had pneumonia or other medical conditions causing the hyperpyrexia, the patient is referred to the general
medical ward or to the medical outpatients. If a patient is
diagnosed as a case of heat exhaustion he would be referred
to the heat exhaustion ward.

When a case is diagnosed as heat stroke, the patient is
immediately transferred to the HSTC where diagnosis is confirmed and management started. The procedures followed at
the HSTC have been outlined by Khogali *et al.* (1983).

The HSTC directly receives cases diagnosed as heat stroke,
from the health centres or medical missions or brought
directly by the ambulance in critical stages.

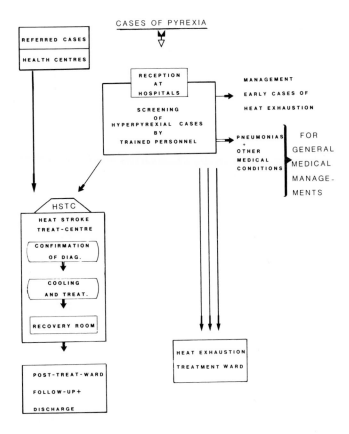

*Fig. 4. Flow Chart of the Management Plan for Cases
of Heat Illness.*

Management at Health Centres

Beginning in 1982, the start of the hot cycles, every health centre was modified to accommodate two small 6-bedded wards for treatment of heat exhaustion – one for males and the other for females. Thus, on screening at the health centre, all cases of heat exhaustion could be managed on the spot.

On diagnosing a case of heat stroke, first aid measures which are life saving, are immediately instituted (Khogali, 1982). The patient is then transported to the nearest HSTC under strict instructions to maintain the airways clear.

Discussion

The heat disorders seen at the pilgrimage are more varied and complicated than disorders reported elsewhere (Ellis, 1976). The multi-nationalities and the underlying chronic and endemic diseases prevalent among many of the pilgrims complicate the picture. Added to this is the sudden influx of one million from outside Saudi Arabia and another million from inside into the Holy Areas.

The plan of management adopted proved to be very successful in that it could help in diagnosing heat stroke cases at the earliest possible time on arrival, and in detection of early cases of heat exhaustion. This is a major problem and its early detection resulted in reducing the number of heat stroke cases. The results of management of heat exhaustion are rewarding, with achievement of a 100% recovery (Al Dabbag et al., 1983).

The major shortcoming of the plan is the difficulty in early detection of cases before they are brought to hospital. To overcome this, very close co-ordination and co-operation is needed between the Mutawif (guides of the pilgrim), the medical missions and the ambulance services. A detailed plan of operation has been suggested to be implemented during the next Hajj.

Acknowledgements

We are very grateful to the Minister of Health for his support and to all the medical staff of the Western Region, Saudi Arabia.

REFERENCES

Al Dabbag, M., Khogali, M. & Ghallab, M. (1983). Clinical
picture and management of heat exhaustion. *In* "Heat
Stroke and Temperature Regulation" (M. Khogali & J.R.S.
Hales, eds), pp. 171–177. Academic Press, Sydney.
Ellis, F.P. (1976). Heat illness. *Trans. Roy. Soc. Trop.
Med. & Hyg. 70*, 402–425.
Khogali, M. (1982). Heat disorders with special reference to
Makkah Pilgrimage. Monograph – Ministry of Health.
Kingdom of Saudi Arabia.

4
Basic Concepts and Applied Aspects of Body Temperature Regulation

John Bligh

Institute of Arctic Biology
University of Alaska
Fairbanks, Alaska

Understandings of the nature of the body temperature set-point determinant, and of the central nervous interface, are essential to any real understanding of the central events and circumstances which can shift or disturb body temperature, or impair the ability to cope with imposed thermal stress. Current concepts of the nature of the set-point are reviewed. Most likely the set-point determinant resides in the activity/temperature characteristics of the deep-body temperature sensors. The neuronal organization at the central interface between temperature sensors and thermoregulatory effectors may have no special properties, but consist essentially of pathways from warm sensors to heat loss effectors and from cold sensors to heat production effectors, with crossing inhibition between them, and the convergence onto these pathways of various excitatory and inhibitory influences from elsewhere in the brain. The effect of these converging influences, which include those from peripheral temperature sensors, is to vary the set-point. Thus any induced activity of the brain caused by exogenous or endogenous disturbances is likely to influence the set-point of body temperature. Many pathological or pharmacological disturbances to central synaptic events may also change or disrupt temperature regulation.

HEAT STROKE AND
TEMPERATURE REGULATION
ISBN 0 12 406180 X

41

This contribution to the Workshop proceedings is not intended to be either an authoritative (e.g. fully documented) or a comprehensive account of mammalian temperature regulation *per se*, or of the various environmental, physiological, pathological and pharmacological disturbances which may directly or indirectly impair the ability of the body to sustain body temperature during exogenous and endogenous heat stress. The intention is that the presentation of a simplified model of the thermoregulatory system, will provide a framework which can be used in considerations of the various details of thermal functions and dysfunctions presented by other contributors.

PROCESSES OF HOMEOTHERMY PARTICULARLY

The processes of homeothermy involve both warm- and cold-sensors, variously located in the body, and two distinct and opposing categories of thermoregulatory correction effectors – those which vary heat production, and those which vary heat loss (Fig. 1).

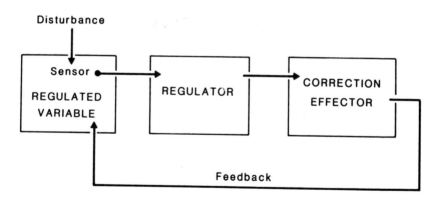

Fig. 1. A schematic representation of mammalian temperature regulation in which the nature of the central nervous interface between the sensors and the effectors remains undetermined.

Temperature sensors are located in the preoptic/anterior hypothalmic region of the midbrain (PO/AH), the medulla, the thoracic spinal cord, the outer surface structures, and perhaps elsewhere more sparsely. The principal central nervous connections between temperature sensors and thermo-regulatory effectors are undoubtedly at the level of the hypothalamus, but probably not exclusively so, since destruction of this part of the brain does not wholly oblit-erate the responses to thermal disturbances. Much evidence, however, clearly implies the convergence of the afferent pathways from temperature sensors elsewhere in the body with those in the PO/AH, such that they exert common and additive actions on the thermoregulatory effectors.

In Fig. 1, the behavioural processes, and the autonomic thermoregulatory effector processes, are jointly represented by the boxes labelled "heat production effectors" and "heat loss effectors". The temperature of the venous blood drain-ing from tissues involved in heat production and heat loss reflects these activities. These thermally-distinct streams of venous blood become well mixed in the heart and lungs, and the temperature of the arterial blood issuing from the left heart into the aorta is thus quickly influenced by alterations in the rates of heat production and heat loss. Since the arterial blood is then distributed throughout the body, it acts as the thermal feedback pathway to the temperature sensors, wherever they may be. There is nothing novel or controversial in Fig. 1. It is simply a preferred way of summarizing the basic structures and their relationships.

A CENTRAL REGULATOR OF BODY TEMPERATURE

A dominant question in central neurology is "On what basis does the CNS interpret the information received from disturbance sensors, and respond with appropriate instructions to the correction effectors?" Except in general terms of neuronal properties and relationships, this is poorly under-stood generally and is a controversial issue in the study of temperature regulation. One assumption is that somehow the central nervous interface not only connects the variously-located warm- and cold-sensors to the various behavioural and autonomic thermoregulatory effectors, but is somehow related to a centrally-located set-point determinant, which is variously supposed to be based upon temperature-insensitive continuously-firing neurones (Hardy, 1969); a biochemical

gating influence on the thermosensor to thermoregulatory
pathways (Feldberg & Myers, 1963);   or an ionic gating
influence on those pathways (Myers & Veale, 1970). There is
no compelling evidence in support of any of these concepts,
but a variety of observations can be, and have been, interpreted as supportive of each of them.

ALTERNATIVE THEORIES ON THE NATURE OF THE REGULATOR

    In recent years, however, doubt has been expressed about
whether the set-point determinant necessarily resides within
the CNS on the pathways connecting temperature sensors to
thermoregulatory effectors, and even whether it is necessary
to postulate a set-point determinant of any sort anywhere
(Houdas *et al.*, 1973; Werner, 1980). Some of these alternative theories deny the very existence of true thermoregulation, but if the level of body temperature were to be
simply the consequence of dynamic balance between the
processes of heat production and heat loss, body temperature
would surely be much more variable than it is. It is, however, evident that changes in body temperatures can result in
the activation of appropriate thermoregulatory effectors, and
that a particular body temperature is thereby actively
defended. Clearly, temperature sensor/thermoregulatory
effector relations are involved in the defence of a more-or-
less fixed level of core temperature. Something, therefore
is somehow acting as a set-point determinant. It need not
be, however, a particular property of a neuronal structure
lying within the CNS between the afferent pathways from
temperature sensors, and the efferent pathways to the thermoregulatory effectors. It could be invested in the genetic-
ally-fixed activity/temperature properties of the warm- and
cold-sensors (Vendrik, 1959): opposing activity/temperature
profiles of the warm- and cold-sensors would almost inevitably
function as a set-point determinant if they modulate the
intensities of thermoregulatory heat production and heat loss
effectors in opposing ways.
    Many subsequent studies of the electrical activities of
cutaneous and central neuronal structures considered to be
sensors, indicate that while any one pair of warm-sensors and
cold-sensors may not have the neat overlap of opposing
activity/temperature profiles envisaged by Vendrik (1959), it
is possible, that the means of many temperature sensors
activity/temperature profiles at a particular site, have something close to the Vendrik relationship. Since it is arguable

that there is more evidence for this concept than there is
for any of the central reference theories, there would seem
to be no need to hypothesize any functions of the CNS which
are peculiar to the regulation of body temperature other than
those of the centrally located temperature sensors.

THE CENTRAL NERVOUS INTERFACE BETWEEN THERMOSENSORS AND
THERMOREGULATORY EFFECTORS

One basic pattern of the neuronal arrangement of the
central nervous interface between temperature sensors and
thermoregulatory effectors (Fig. 2a) seems to be emerging.
This consists of two sensor-to-effector pathways, from the
warm sensors to heat loss effectors, and from cold sensors to
heat production effectors, with crossing inhibitory influences between them. This was first proposed on the basis
of the disturbance/response relations of dogs (Hammel, 1965)
and later to express the influences of deep body and skin
temperatures on the autonomic thermoregulatory activities
of man (Wyndham & Atkins, 1968), with the additional proposition that the pathways from peripheral and central temper-
ature-sensors converge. A similar model was used, quite
independently, to express the effects on thermoregulation in
sheep of intracerebro-ventricular injections of various
putative synaptic transmitter substances (Bligh *et al.*, 1971).
Many further such studies have required no change in the
basic concept (Bligh, 1979).
    These three very similar presentations bear striking
resemblances to that used by Sherrington (1906) to express
the reciprocally inhibiting influences between the neural
pathways by which opposing flexor and extensor muscles are
instructed (Fig. 2b).
    From Sherrington's study of the integrative action of the
nervous system, it is apparent that every sensor-to-effector
pathway contains at least one central synapse at which the
stimulus/response relationship is modified by excitatory and
inhibitory influences derived from other events in the CNS,
and that this highly organized exchange of influences is
basic to the integrative functions of the CNS. Thus no
sensor-to-effector pathway through the CNS is invariably
fixed. The crossing inhibition between two opposing effector
functions (Fig. 2b) can be considered as a particular case of
this synaptic modulation, as can the convergance of the pathways from peripheral and central temperature sensors

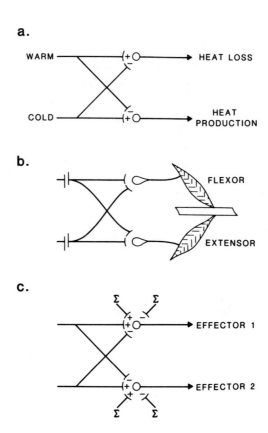

Fig. 2.   (a) The basic concept of the relations between temperature sensors and thermoregulatory effectors according to Hammel (1965), Wyndham & Atkins (1968) and Bligh, Cottle & Maskrey (1971).   (b) The crossing inhibition between the efferent pathways to antagonistic flexor and extensor muscles (from Sherrington, 1906).   (c) An interpretation of the Sherringtonian concepts of basic central neuronal organization, where (Σ) = summed excitatory (+) and inhibitory (−) influences acting on antagonistic effector pathways, between which there is also crossing inhibition (see Bligh, 1984).

indicated by Wyndham & Atkins (1968). These Sherringtonian principles of central neurology can be summarized as in Fig. 2c.

Thus the basic relations between temperature sensors and thermoregulatory effectors, derivable independently from

observed relations between thermal disturbances and thermoregulatory responses, and from studies of the effects of
central synaptic interferences on thermoregulatory events,
are virtually identical to the classical template of central
nervous organization which can be constructed from
Sherrington's consideration of the neural events at the level
of the spinal cord. This basic pattern of the possible
relations between temperature sensors and thermoregulatory
effectors fits with the proposition, based upon Vendrik's
principle, that there is no need to hypothesize special
features of the central nervous interface to account for the
set-point of body temperature. Furthermore, the synaptic
influences upon the core temperature sensor-to-thermoregulatory effector pathways, both from peripheral temper-
ature-sensors, and from elsewhere in the CNS and relating to
non-thermal events, would be expected to vary the effective
set-point of temperature regulation by varying the core
temperature sensor-to-thermoregulatory effector relations;
such changes in 'set-point' occur.

A TENTATIVE CONCLUSION ABOUT THE NATURE OF MAMMALIAN
TEMPERATURE REGULATION

On the basis of the above considerations, it is possible
that i) the set-point determinant resides in the genetically
-fixed activity/temperature characteristics of the warm- and
cold-sensors; ii) the central nervous interface between the
afferent pathways from the temperature sensors and the
efferent pathways to the thermoregulatory effectors is
basically no different than that which exists generally in
the CNS, and iii) the variability of the effective set-point
of temperature regulation despite the fixation of the
properties of the temperature sensors, is the consequence of
the synaptic convergence of other influences.
When this arrangement of the possible central connections
between the warm- and cold-sensors and the opposing heat loss
and heat production effectors (Fig. 2c), is inserted into
Fig. 1, it provides what appears to be a workable basis for
the processes of temperature regulation (Fig. 3). In this
system the crossing inhibition does not create the set-point,
but it does create the null-point or null-zone at, or between,
the temperature thresholds for thermoregulatory heat loss and
heat production effector activities. The converging
excitatory and inhibitory influences, including those from
peripheral temperature sensors, modify the relations between

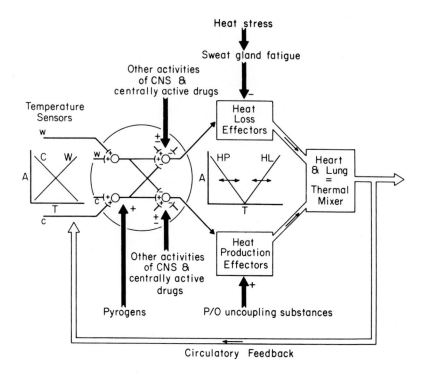

Fig. 3. An amalgamation of Figs 1 and 2c to show the possible basic nature of the central nervous interface between cold (C) and warm (W) sensors and the heat production (HP) and heat loss (HL) effectors. The inset plots are of the activity (A)/temperature (T) relations of the sensors (left) and effectors (right). The heavy arrows indicate the actions and points of action of sundry interferences with thermoregulatory processes.

the activities of the core temperature sensors and the thermoregulatory effectors, thereby causing temporary changes in the set-point or set-range of the regulated body temperature.

Even if valid, obviously this concept is a gross over-simplification, but no aspect of it is novel: there must be synaptically-gated pathways through the CNS from the hypo-thalamic temperature sensors to the thermoregulatory effectors, and there must be convergences onto these synapses of other influences including those from extra-hypothalamic temperature sensors.

THE PATHOLOGY OF THERMOREGULATION

The hyperthermia of exercise or of exposure to extreme heat may be wholly attributable to the saturation of fully functional heat loss effector processes, but abnormalities in the processes of heat production or of heat loss, or in the neural processes by which they are controlled, may be contributory or even dominant factors. These may include i) general metabolic abnormalities (e.g. hyperthyroid activity); ii) central metabolic abnormalities particularly in relation to synaptic events (e.g. Parkinson's disease); iii) a febrile response to infection; iv) destruction of pathways from temperature sensors or to thermoregulatory effectors (e.g. disseminated demyelination); v) damage to the central nervous interface (oncologic or haemorrhagic destruction of brain tissue); vi) toxaemia acting centrally or extra-centrally (e.g. accidental exposure to organophosphates); vii) disturbances to central events directly or indirectly affecting temperature regulation consequent upon the prescription of centrally-active drugs for other neurological or psychological conditions; and viii) similar disturbances due to the self-administration of centrally-active substances. Since both central and extra-central perturbations need to be considered, a model of temperature regulation in which the central nervous interface between the afferent pathways from temperature sensors and the efferent pathways to thermoregulatory effectors is expressible in terms of a pattern of neuronal interactions is helpful, because it allows the representation of disturbances to central functions in terms of synaptic interactions between neurones. Central synapses can no longer be considered in simple terms of a transmitter substance released by one cell and acting upon another, and seldom dysfunctioning to such an extent as to impair whole-body functions significantly. On the contrary, synaptic biochemistry and physiology is exceedingly complex, and dysfunction of central synaptic events may be at the basis of many gross physiological abnormalities.

Recognized synaptic dysfunctions include: a) autoimmune disturbances to central synaptic function comparable to the neuromuscular dysfunction in myasthenia gravis; b) metabolic defects in transmitter synthesis such as is apparent in Parkinson's disease; c) dietary deficiencies of essential precursors of transmitter substances; and d) disturbances to synaptic events induced by centrally-active drugs which may

be medicinal (e.g. prescribed tranquillizers), socially
acceptable (e.g. tea, coffee, alcohol) or illicit (e.g. self
administered opiates).

Clearly, synaptic inadequacies at the central nervous
interface relating to temperature regulation which may become
apparent only when the system is striving to maximize the
responses to extremes of thermal stress, need to be
considered. The growing evidence of feedback of circulatory
hormones onto central nervous cells is another source of
disturbance which might influence temperature regulation
directly or indirectly.

The heavy arrows pointing to the central nervous interface in Fig. 3 are indicative of these many various neuronal
disturbances or dysfunctions which may act upon the temperature sensors-to-thermoregulatory effector pathways, and vary
the balance of the efferent drives to the heat production and
heat loss effectors, and therefore vary the core temperature
at which heat loss equals heat production. There is also a
heavy arrow in Fig. 3 to indicate the apparent point of
action of a pyrogen which raises core temperature by increasing the drive to the heat production effectors while inhibiting the drive to the heat loss effectors.

There is no clear evidence of the effects on temperature
regulation of abnormal activities of central temperature
sensors, but this would be hard to come by because of the
difficulty of distinguishing between the effects of central
disturbances to the sensors, and to the subsequent inter-
neurones at the central sensor-effector interface. Widespread lesions of the skin must modify the thermal information derived from the skin, but on the assumption that core
receptors are dominant as the determinants of core temperature, a gross disturbance to core temperature might not
result even from a substantial deprivation of information
from peripheral temperature sensors. Any effect of cutaneous
sensor deprivation, however, is probably masked by the deprivation of the role of the skin as a physiologically-variable
heat loss structure. The impairment of the cutaneous
vasculature, and of the sweat glands, will interfere with the
control of heat loss by the physical channels of convection,
radiation and evaporation. The failure of the sweat glands
per se or, very occasionally, the genetic absence of sweat
glands, can be a major factor in the inability of an individual to counter a high exogenous or endogenous heat load. This
is also indicated by a heavy arrow in Fig. 3. Another arrow in
Fig. 3 indicates a rise in heat production as a cause of
hyperthermia. This could be due to the additive heat production of exercise during exposure to a high ambient temperature,

but it could also be due to the absorption of a toxic substance such as the organophosphates, which enhance tissue
heat production generally, by uncoupling the processes of
oxidation and phosphorylation.

FINAL COMMENT

This cursory overview of the physiology of temperature
regulation and the potential contributions to its failure
during heat stress indicates the need to consider multiple
causes and pre-disposing conditions.

REFERENCES

Bligh, J. (1979). The central neurology of mammalian thermoregulation. *Neuroscience 4*, 1213-1236.
Bligh, J. (1984). Temperature regulation: a theoretical
consideration incorporating Sherringtonian principles of
central neurology. (Proc. Intl. Symp. Thermoregulatory
Mechanism, Osaka, Japan, 1982). *J. Therm. Biol.*
(in press).
Bligh, J., Cottle, W.H. & Maskrey, M. (1971). Influence of
ambient temperature on the thermoregulatory responses to
5-hydroxytryptamine, noradrenaline and acetylcholine
injected into the lateral cerebral ventricles of sheep,
goats and rabbits. *J. Physiol., Lond. 212*, 377-392.
Feldberg, W. & Myers, R.D. (1963). A new concept of
temperature regulation by amines in the hypothalamus.
Nature, Lond. 200, 1325.
Hammel, H.T. (1965). Neurones and temperature regulation.
In "Physiological Controls and Regulation: (W.S. Yamomoto
& J.R. Brobeck, eds), p.71. Saunders, New York.
Hardy, J.D. (1969). Brain sensors of temperature. Brody
Memorial Lecture VIII. Agric. Exptl. Sta. Spl. Rpt. 103,
University Missouri, Columbia.
Houdas, Y., Savage, A., Bonaventure, M. & Guieu, J.D. (1973).
Modèle de la résponse évaporatoire à l'augmentation de la
charge thermique. *J. Physiol. (Paris) 66*, 137-162.
Myers, R.D. & Veale, W.L. (1970). Body temperature: possible
ionic mechanism in the hypothalamus controlling the set-
point. *Science 170*, 95-97.
Sherrington, C.S. (1906). "The Integrative Action of the
Nervous System". Yale, New Haven.

52   *John Bligh*

Vendrik, A.J.H. (1959).   The regulation of body temperature
    in man.   *Ned. Tijdschr. Geneesk. 103(5)*, 240-244.
Werner, J. (1980).   The concept of regulation for human body
    temperature.   *J. Therm. Biol. 5*, 75-82.
Wyndham, C.H. & Atkins, A.R. (1968).   A physiological scheme
    and mathematical model of temperature regulation in man.
    Pflügers Arch. 303, 14-30.

5
Mechanisms of Human Thermoregulation

T. H. Benzinger

National Bureau of Standards
Washington, DC, USA

*The mechanisms of human thermoregulation are reflex res-
ponses by loss or gain of* heat *to stimuli of* temperature *at
specific sensory organs:* Warm sensors *in the pre-optic
region of the hypothalamus, and* cold sensors *under the surface
of the skin. The skin sensors elicit metabolic overproduction of heat in·response to cold. The central warm sensors
elicit sweat secretion and cutaneous vasodilatation for increased heat loss. The power of these responses compared
with the subtleness of stimulation is extraordinary: A 0.1°C
change of central temperature may double the normal rate of
heat loss or production. Skin cold reception inhibits sweating, and central warm reception inhibits overproduction of
heat. "Isotherm plotting" permitted us to disentangle the
individual mechanisms from the complexity of their concerted
action. However, the primary prerequisites for this work
were two inventions: (1) Direct, gradient-layer calorimetry
for measuring heat losses and body conductance, and (2) tympanic thermometry for measuring the temperature of the central
sensors. Whereas the supreme role of the anterior hypothalamic warm sensors as a "temperature-eye" for physiological
thermoregulation is well established, their role in hyperthermic illness is yet to be defined, by measurements of central temperature in patients.*

There was a time when physiologists believed that all
sensory information for the regulation of temperature origin-
ated from the skin, its receptors for warm and cold. Not only
the conscious sensations of temperature on which the thermoregulatory behaviour depends, but also the unconscious and

unwillful mechanisms of sweating and cutaneous vasodilatation
for heat loss and shivering for increased metabolic heat production were thought to be elicited by warm or cold receptors
of the skin. They reached by afferent pathways, nerve centres
from which they were relayed out to the effector organs,
sweat glands, blood vessels or shivering muscles.

Since it was known that central temperature, too, had a
definite role that affected the thermoregulatory response, it
was believed that the relevant centres when warmed or cooled
were either facilitating or inhibiting the transmission of
impulses from the sensory into the effector pathways. Central
temperature thus modified the "sensitivity" of the centres to
the incoming messages from the skin. Doubts about these concepts were expressed in 1958 by Sir G. Pickering (Pickering,
1958), to quote:

> *"they think that the evidence that exists for some
> central control can be explained by the effect of
> temperature upon the relaying and conducting
> functions of these centers".... "The evidence
> collected here leads me to think that this is not
> a correct assessment of the situation.".*

Doubts and uncertainty persisted. As late as in the
Physiological Review of 1961 (Hardy, 1961) knowledge of
temperature regulation was summarized as follows:

> *"In summary there is presented in this section a
> qualitative description of a concept of action of
> the regulating systems".... "The temperature
> receptor systems of the skin, hypothalamus, and
> other body areas have similar properties and
> participate in this regulation".... "The combined
> effects of the cold and warmth receptors prescribe
> in some as yet undetermined additive manner the
> action of the thermoregulation"* and *"Exact pro-
> portions for the effects of peripheral and central
> receptors cannot be evaluated at this time".*

A total lack of quantitative understanding was thus evident concerning a physiological mechanism of incredible power
and precision. The central temperature of man is maintained
within a narrow range of $\pm 0.5°C$ against external or internal
loads up to four times or more of a normal rate of heat production or heat loss. How is this power attained? How is
this precision achieved? The stimuli of temperature must be
subtle indeed. The responses by loss or over-production of
heat must be extremely strong and swift. For both the stimuli
and the responses sensitive, accurate and rapidly responding
methods of measurement were needed.

METHODS OF MEASUREMENT

Fortunately since 1938 (Benzinger, 1938), persistent efforts in our laboratory had been made to develop quantitative
methods of investigation. First using continuous gas analysis
invented by Hermann Rein in Göttingen (Rein, 1933) a way was
found for continuous direct recording, at low inertial dis-
tortion, of the rates of oxygen consumption and carbon dioxide
production. The subject inhaled from and exhaled into a
constant flow of air of a velocity just above maximum respiratory volume. Downstream, a sample of the diluted expired
air was continuously withdrawn for gas analysis. The analyser
recorded oxygen consumption rate against time, in steady
states or during rapid transients with minimal inertial dis-
tortion. Only many years later, in 1960, was this method
used again in our experiments in conjunction with direct
gradient layer calorimetry or during immersion of subjects in
cold or warm water baths. The aim was to correlate the
metabolic rates observed to central or cutaneous stimuli of
temperature as will be shown in successive sections of this
paper.

The next step was initiated during World War II at
Rechlin, Germany, with a pilot project called "Snowhite", a
first attempt at direct, gradient-layer calorimetry of human
subjects. This project could not be carried on during war-
time in Germany. It was resumed in 1947 by the US Navy at
the National Naval Medical Center in Bethesda.

Workshop and engineering requirements exceeded the
facilities of a research institute. Construction was therefore carried out at the Laboratory of the American Society of
Heating and Ventilating Engineers in Cleveland, Ohio, during
the years from 1947 to 1957 (Benzinger *et al.*, 1958).

Human and animal calorimetry for applications in
Nutrition Science or thermophysiology had a tradition of high
accuracy of measurement. The intended measurements of heat
loss rate should therefore be correct within ±1 or 2%. A
half-response time of less than one minute was desired. To
achieve these goals the principle of gradient layer calorimetry was conceived and developed.

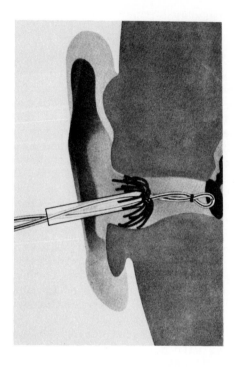

Fig. 1. A tympanic temperature sensor as used in the
experiments reported in this paper.

PHYSICAL THERMOREGULATION AGAINST OVERHEATING

When the human gradient calorimeter became ready for
operation in 1957 and when it had the desired performance
characteristics, the intent was first to measure heat loss
rate in warm environments and to relate the rates of sweating
heat loss to the temperature of the skin, in accordance with
the theory that warm reception at the skin would cause sweating.   The relations found were completely erratic and sense-
less.

Next we attempted to find reproducible relations to
central temperature measured in the *rectum* with the same negative result, for different reasons.   As we found later, the
skin is not the origin of driving impulses for sweating, and
rectal measurements do not truly represent the central temperature of man.   After the introduction of tympanic thermometry

(Fig. 1) for our experiments with the human calorimeter the results became immediately clear-cut and consistent.

In warm environments, where the skin temperature of resting or working subjects is above the skin cold receptor threshold, human subjects are sweating if their central temperature is high enough. However, the rate of sweating is zero as long as central (tympanic) temperature is lower than the "set-point", an individual characteristic of the subject, close to 37°C. With temperature rising *beyond* the set-point, the rate of sweating increases *linearly* with the upward deviation, regardless of skin temperature (Fig. 2). It followed that the hypothalamic centre was not a relay station for warm impulses from the skin. It was recognized to be an independent sensory organ with first neurons for temperature-reception, a temperature eye, so to speak, a brother to the retina, the receptor organ for photons (Benzinger, 1959).

Shortly after our prediction, Teruo Nakayama (Nakayama & Eisenman, 1961) discovered thermosensitive firing neurons in the preoptic region of the hypothalamus, and Hellon (1967) observed their "set-point" characteristics. The set-point is a sensory receptor threshold.

Using the capability of our calorimeter for measuring "conductance" of the skin and thereby, vasodilatation, we soon found that vasodilatation rises in the same fashion beginning at the same set-point. This mechanism makes it possible for body heat to be transported from interior to skin at the same rate as it flows from the skin to the environment, mainly by sweat evaporation. If this were not so, the sweat could not evaporate and would drop off without fulfilling its function.

A COMPLICATION AND ITS RESOLUTION

One reason why the straight relationship between the sweat rate and the central temperature was not discovered earlier, was *inhibition* of sweating when cold reception at the skin occurs while warmth reception in the brain takes place (Benzinger, 1961). This paradoxical situation is not infrequent in everyday life. The athelete in a cold environment on the bicycle ergometer was a favoured object in classical temperature physiology.

Such complex situations can be disentangled by plotting sweat rates against central temperature and drawing isotherms through measured points of equal skin temperature, or by plotting sweat rates against skin temperatures with isotherms

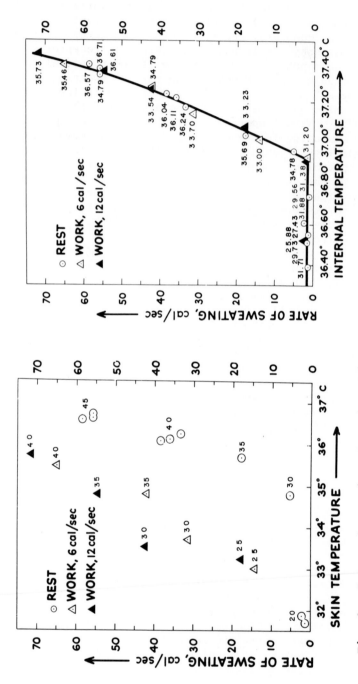

Fig. 2.   Rates of sweating plotted against skin temperatures (left) and tympanic temperatures (right).

drawn through points of equal measured central temperature. This method has enabled us to separate the two conflicting mechanisms in man's "physical" thermoregulation against over-heating. It was also instrumental later in the investigation of "chemical" thermoregulation by over-production of heat as will be shown below.

RELEVANCE TO THE PILGRIMAGE

The finding that the temperature of the brain stem can be conveniently and reliably measured and that it is the very stimulus to which the two mechanisms of sweating and vaso-dilatation respond with quantitative precision has fundamental importance for research on hyperthermic illness and the dangers of the pilgrimage.

Every defense mechanism of the human body has its limits beyond which it cannot maintain or restore the vital homeo-stasis. When sweating and vasodilatation reach a maximum, and central temperature, nonetheless, keeps rising under external or internal heat loads, the central sensor itself must suffer from the heat and will no longer function. A vicious cycle of overloading and diminishing response will set in; finally, a point of no return will be reached.

Since rectal temperature bears no relation to the temper-ature of the brain stem, which is so far away anatomically, it is not surprising that both well-being or a near-fatal condition of a patient have been observed at one and the same *rectal* temperature. Rectal measurements are misleading. Only central, say tympanic, measurements can be expected to be consistent with the clinical condition of the patient, with the severity of his or her hyperthermic illness.

Once the severity has been diagnosed, treatment will be initiated. Here again, central, not rectal temperature will reflect the efficiency of the method of choice: Cooling by evaporation from atomized water envelopment of the patient according to Weiner & Khogali (1980). Such monitoring has been carried out before, on divers recovering from *hypo*-thermia. It has shown central and rectal temperatures not only widely divergent but also moving in opposite directions (Benzinger, 1969). We fully and confidently expect that monitoring pilgrims' hyperthermic condition by measurements of central temperature will permit a precise, quantitative diagnosis of the severity of the illness and a clear picture of the effectiveness of the treatment.

CHEMICAL THERMOREGULATION

The title of this paper calls for a description of human temperature regulation, and this includes the chemical defense against body cooling. Although chemical thermoregulation is not directly relevant to our main theme, the pilgrimage in a hot desert, a complete treatment will bring the overriding importance of central temperature again into focus.

Central temperature, not skin temperature is the regulated norm, in cold as well as in warm environments. To be sure, the metabolic over-production of heat is *elicited* by cold receptors of the skin, but its magnitude is strictly controlled by *central warm inhibition* as we shall see (Benzinger, 1961). How does this dual mechanism work, and what is the experimental proof of its function?

The origin of the metabolic response from cold receptors at the skin is proven by the fact that the patterns of oxygen consumption rates imitate in every detail the patterns of the firing rates of the cold receptors of the skin. In transients, a salient feature is the over-shooting response to a sudden *drop* in skin temperature and over-shooting inhibition upon a sudden *rise* in skin temperature.

In steady states the method of isotherm plotting reveals the response by over-production of heat as a result of peripheral *cold excitation* and central *warm inhibition*. The inhibition becomes *total* when central temperature rises beyond the set-point at which the responses of sweating and vasodilatation begin. When central temperature falls, while the skin is cold (say 20°C for maximal cold receptor firing) the metabolic response to these impulses is gradually released from the central warm inhibition. It can reach an intensity of more than four times a resting rate of human metabolic heat production (Fig. 3).

THE CENTRAL WARM INHIBITION OF THE METABOLIC RESPONSE TO COLD

Since the metabolic response is elicited by *cold* receptors (skin) which increase their firing rates on *cooling*, the central inhibition must be caused by *warm* receptors which increase their firing rates on *warming*. These central warm receptors are sharply distinguished from those that elicit the responses of sweating and vasodilatation. They begin to fire at temperatures more than one degree Celsius *below* the sweating set-point. When temperature rises *to* the set-point

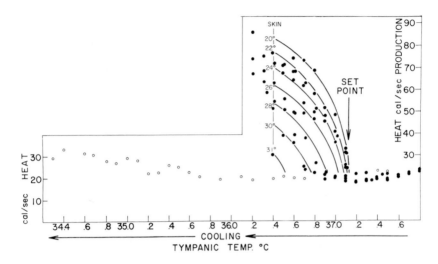

*Fig. 3.   Central cold reception.   Circles at left show on
a paraplegic patient the leftover metabolic response to
central cold reception.   Man with intact nervous system (dots
at right) responds to driving impulses from skin cold receptors released from central warm inhibition at the set-point,
37.1°C.   In the paraplegic patient, oxygen consumption begins
to rise only below 35.7°C central temperature, to a maximum
of +50% BMR, with a sensitivity of ~+10 (cal sec⁻¹ deg⁻¹).
In man with intact nervous system (dots) oxygen consumption
rises from the set-point, 37.1°C, with an initial intensity
of ~+200 (cal sec⁻¹ deg⁻¹), to sustained metabolic rates of
+300% BMR (not measured in the range below 36°C central temperature of this figure).*

the inhibition of the over-production of heat becomes *total*,
regardless even of the most intense skin cold receptor firing.
Warm sensing neurons with matching characteristics ("Type A-
neurons") have been discovered in the preoptic centre
(Hellon, 1967).

It has been suggested by others that the roles of skin
and central receptors in chemical thermoregulation are
"additive", that they are both elicited by cold sensing
neurons. If this were so, it should be possible for either
mechanism in the absence of the other to elicit the observed
over-production of heat in response to cold. Experiments
have shown this not to be the case. Even maximal excitation
of skin cold receptors cannot elicit the metabolic response

as long as central temperature is above the sweating set-point. Even low central temperature, down to 35.8°C, does not elicit the metabolic response when cold receptors of the skin are not stimulated (Fig. 3). The metabolic action can be explained *only* by antagonistic action, peripheral cold excitation and central warm inhibition.

CONCLUSIONS FOR THE PILGRIMAGE

The central "temperature eye" as it elicits sweating and vasodilatation in warm environments and during hard work, and as it regulates the metabolic over-production of heat in response to cold reception at the skin by warm inhibition, is totally in control of all thermoregulatory mechanisms in the life of man. These facts are fundamental, particularly to our perception of hyperthermic illness, heat in the desert, and the pilgrimage. The crucial method of investigation is the measurement of *central* temperature.

THE BEHAVIOURAL REGULATION OF HUMAN TEMPERATURE

The autonomic regulation of human central temperature described in the preceding sections is automatic, unconscious and unwilful. It excels in extraordinary precision. Its power is, however, limited to approximately four resting metabolic rates.

Another set of mechanisms, conscious and wilful, is also available to man, viz., behaviour. It is less accurate, but almost unlimited in its power to cope with any load, on land, under water, in the atmosphere and in space. If the pilgrims could travel in air-conditioned vehicles, there would be no demands on autonomic thermoregulation and no hyperthermic illness. Technology would have eliminated our problem. Even without technology as one kind of human behaviour there are other wilful and conscious actions that can reduce environmental or internal heat loads: moving out of the sun, into the shade, seeking shelter, avoiding exertion, donning or shedding clothes. All of such actions require a sensory component, a conscious perception of temperature, warm or cold. What are the sensory sources of human thermoregulatory behaviour?

Behavioural defense against cold originates from cold receptors of the skin, behavioural defense against overheating originates from central hyperthermia (Benzinger, 1963). Vague perceptions of being unpleasantly warm, with increasing malaise are felt once *central* temperature has moved beyond the set-point for sweating. These sensations can be countermanded by cooling of the skin, but they do not originate from skin warm reception. What is the experimental proof for these statements?

When a human subject, mildly pre-cooled, goes into a 38°C water bath, he or she feels most pleasantly comfortable, although the skin is hotter than it would ever be in a fiery desert. Only very gradually, as *central* temperature rises beyond the set-point for sweating, the bather will begin to feel unpleasantly overheated in stages ranging from mild discomfort to the limit of tolerance and an irresistible desire to get out of this hot bath. Central, not skin temperature is the source of the complaint.

BEHAVIOURAL DEFENSE AGAINST COLD

In constrast, when a bath of neutral temperature is gradually made cooler, the subject will have a first sensation of cooling when bath temperature moves below the skin cold receptor threshold of 35 to 34°C. The threshold of this sensation was found to be *independent*, over a wide range, *of the central (tympanic) temperature* of the subject. This proves that there is *no* contribution to feeling cold, from central cold sensors, at least not until a marked hypothermia is established, below 35.8°C, and even then the response is weak, (50% in a paraplegic, as compared with 400% in an individual with intact sensory pathways) (Fig. 3).

In summary, in hot environments, our main topic, not only autonomic but also behavioural thermoregulation originates from *central* temperature which is the objective yardstick for the subjective misery of hyperthermic illness; it ought to be measured and monitored.

REFERENCES

Benzinger, T.H. (1938). Untersuchungen über die Atmung und den Gasstoffwechsel. *Ergeb. Physiol. 40*, 1-52.

Benzinger, T.H. (1959).   On physical heat regulation and the sense of temperature in man. *Proc. Natl Acad. Sci. U.S.* *45*, 645–659.

Benzinger, T.H. (1961).   The diminution of thermoregulatory sweating during cold-reception at the skin. *Proc. Natl Acad. Sci. U.S. 47*, 1683–1688.

Benzinger, T.H. (1963).   Peripheral cold- and central warm-reception, main origins of human thermal discomfort. *Proc. Natl Acad. Sci. U.S. 49*, 832–839.

Benzinger, T.H. (1969).   Heat regulation, homeostasis of central temperature in man. *Physiol. Rev. 49*, 671–759.

Benzinger, T.H., Huebscher, R.G., Minard, D. & Kitzinger, C. (1958).   Human calorimetry by means of the gradient principle. *J. Appl. Physiol. 12*, S1–S24.

Benzinger, T.H., Pratt, A.W. & Kitzinger, C. (1961).   The thermostatic control of human metabolic heat production. *Proc. Natl Acad. Sci. U.S. 47*, 730–739.

Hardy, J.D. (1961).   Physiology of temperature regulation. *Physiol. Rev. 41*, 521–606.

Hellon, R.F. (1967).   Thermal stimulation of hypothalamic neurones in unanaesthetized rabbits. *J. Physiol. 193*, 381–395.

Nakayama, T., Eisenman, J.S. & Hardy, J.D. (1961).   Recording of activity in the anterior hypothalamus during periods of local heating. *Science 134*, 560–561.

Pickering, G. (1958).   Regulation of body temperature in health and disease. *Lancet i*, 1–9.

Rein, H. (1933).   Ein Gaswechselschreiber. *Arch. Expt. Pathol. Pharmacol. 171*, 363–402.

Weiner, J.S. & Khogali, M. (1980).   A physiological body cooling unit for treatment of heat stroke. *Lancet ii*, 276–278.

6
Set-Point Shift in Thermoregulatory Adaptation and Heat Stroke*

Moneim Attia and M. Khogali

Department of Community Medicine
Faculty of Medicine
University of Kuwait
Kuwait

Thermal adaptation in homeotherms is the modification of normal thermoregulatory response. Adaptive modifications may occur in the CNS, effectors or in peripheral receptors. CNS modifications are characterized by a shift in the threshold of response i.e. the thermoregulatory set-point.

Long term medical observations of individuals suffering from functional or anatomical loss of extremities revealed that in spinal cord transections a state of partial poikilo-thermia is observed some years after the injury. Paraplegics tend to raise their thermoregulatory set-point in heat and lower it in the cold. Cases of congenital maldevelopment of limbs are characterized by a raised core temperature before the age of 12 years and later a lower core temperature when compared against non-disabled children. However, adult amputees fail to adapt and thus suffer from excessive sweating on the trunk to compensate for the loss of extremities.

In this paper it is speculated that during heat stroke the body defends itself by an upward shift of the thermoregulatory set-point. This is a CNS intervention when excessive skin vasodilatation causes blood supply to vital organs to decrease to dangerous levels. The upward shift of the

Supported in part by Research Council Grant No. MC 009 and Seed Money Grant No. MDC046, Kuwait University and Kuwait Foundation for Advancement of Science.

65

set-point decreases the demand for cutaneous circulation resulting in the improvement of the blood supply to the vital organs. Evidence supporting this hypothesis is based on clinical observations of patients suffering from heat stroke and on experimentally induced heat stroke on sheep. In heat stroke patients it was observed that the thresholds for sweating and shivering were sizably raised, whereas in sheep suffering from heat stroke the thresholds for panting and vasoconstriction were raised.

Thus, during heat stroke a situation similar to that with fever prevails. In fever, body temperature is closely regulated at a higher level than normal. In heat stroke, the thermoregulatory system fails and the demand of vital organs for blood overrides the thermoregulatory cutaneous demand.

Thermal adaptation in homeotherms can be considered as modifications of normal thermoregulatory responses (Hensel *et al.*, 1982). In most cases, modifications in normal thermal strain are brought about by a continuous or prolonged periodic endogenous and/or exogenous thermal stress. The time requirement differs according to the different kinds of adaptive changes or processes; ranging from seconds in receptor adaptation to many generations in selection of genetically adapted types (Hildebrandt, 1982). Thermoregulatory adaptive modifications result in a tolerance increase, economy of regulation and normalization of functional efficiency. They occur in the central nervous system (CNS), e.g., set-point shift, in the effectors, e.g., capacity change of sweat glands and in the peripheral receptors (Hensel *et al.*, 1982).

This presentation shall be confined to some aspects of CNS modifications in response to a total or partial failure of the thermoregulatory system, such as in heat stroke or in functional and anatomical loss of extremities, respectively.

TEMPERATURE REGULATION

The objective of temperature regulation in man is to maintain body temperature through the processes of heat production and heat loss, at a stable level despite fluctuations in ambient thermal conditions. Temperature regulation has no specific organ of its own, except sweat glands, but it makes use of other organs and behavioural means. The total temperature regulatory response (R) is the result of both behavioural (B) and autonomic (A) reactions:

$$R = B + A$$

Thermal balance is achieved by the total temperature regulatory response (R), in such a way as to minimize autonomic responses (A). In other words, behavioural responses tend to modify the need for autonomic temperature regulatory responses (Cabanac, 1972; Hardy, 1972; Hensel *et al.*, 1973).

TEMPERATURE CIRCADIAN RHYTHM

The set function of the human temperature regulatory system is complex. The value of the set function at 5 different times of the 24-hour day were estimated by Cabanac *et al.* (1976) using a behavioural indicator. Using the same techniques, Attia *et al.* (1980) exposed a group of young men in a climatic chamber at 8 different times of the day and were able to write the following function, which attempted to describe the circadian variations of the set signal in man:

$$T_{set} = 37.1 + 0.3\cos 15(t-16)$$

where T_{set} = the value of the set function at any point in time of the 24-hour day, i.e., the set-point.

and   t = time of day in hours.

Fig. 1 shows rectal temperature and rectal temperature set-point of a group of young men plotted against time of day. It can easily be observed that the prevailing rectal temperature during rest in an atmosphere of thermal neutrality approximately attained the same values as the set function throughout the 24-hour day. There is a pronounced nycthemeral fluctuation with the rising phase starting in the early morning and the falling phase in the early afternoon. Since the 24-hour rhythm of body temperature is present in a temperature range where the control actions are near zero, this variation is obviously due to a shift in set-point.

MENSTRUAL CYCLE

Cunningham & Cabanac (1971), tested a group of female subjects using local thermal pleasantness rating, and concluded that during the menstrual cycle a shift in the set-point was evident. According to these authors the observed

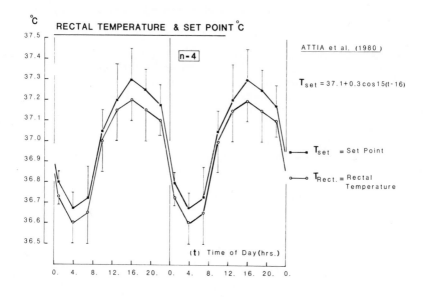

Fig. 1.   Mean ± s.e. of rectal temperature and thermo-regulatory set-point, plotted twice against time of day. From: Attia et al. (1980), courtesy of Springer Verlag.

0.5°C difference in the pre- and post-ovulatory rectal temperature is due to a displacement of set-point.

SPINAL CORD TRANSECTIONS

It is known that paraplegics and tetraplegics suffer from a condition of partial poikilothermia. They show a lower core temperature in the cold and higher core temperature in the heat when compared against physically non-disabled. Extreme cases were reported with core temperatures as high as 42.8°C resulting from impaction of faeces or irritation by a cathartic (Paget, 1885), and 42.0°C observed in a patient with a severe incomplete lesion below C4 (Guttmann, 1976); and as low as 30.8°C observed in a patient three days after

an injury from the level of C6 (Pledger, 1962), and 29.0°C
recorded at the Stoke Mandeville Centre in a tetraplegic
(Guttmann, 1976).

Most of the contributors confirmed that partial failure
to maintain a constant core temperature independent of
fluctuations in ambient temperature is due to the lack of an
efficient system of vasoconstriction, vasodilatation, and
sweating in the insentient portion of the body. Observations
which may indicate that partial poikilothermia is associated
with a shift in thermoregulatory set-point are few (Downey *et
al.*, 1969, 1976; Attia & Engel, 1983).

Fig. 2 is a summary of 72 passive thermal exposures of a
group of paraplegics and a control group of physically non-
disabled subjects. Both groups were exposed to a set of
ambient temperatures ranging from 15°C to 40°C in a climatic
chamber. Thermal pleasantness rating was used to estimate
deviations from thermoregulatory set-point (Cabanac, 1969;
Attia & Engel, 1982). The so-called rating/stimulus slope is
an indicator of the thermal state of the subject. Positive
slopes indicate hypothermal conditions, negative slopes
indicate hyperthermia, whereas, zero slopes indicate normothermia or thermal neutrality (Attia & Engel, 1981).

Fig. 2-B confirms the well known state of partial
poikilothermia in spinal cord transections. The control
group maintained a fairly constant rectal temperature during
all exposures, whereas paraplegics had lower core temperature
in the cold and higher core temperature in the heat. The
rating/stimulus slope for the control group (Fig. 2-A) is a
sigmoid curve showing a sharp point of intersection with the
zero line. Paraplegics, however, show rating/stimulus slopes
closer to the zero line when compared against the control
group (Fig. 2-A). These results can be interpreted by
considering the load error (T_{rect} - T_{set})°C for both groups.
It has previously been shown (Attia & Engel, 1981) that when
the rating/stimulus slope curve intersects the zero line, the
prevailing core temperature is equal to the thermoregulatory
set-point. In the spinal cord transections the load error
seems to be less varying with ambient thermal conditions than
in the control group. A possible way for paraplegics to keep
the load error under control is by shifting their thermoregulatory set-point in the direction of the prevailing core
temperature. This is viewed as a long term temperature
regulatory adaptive change at the level of the CNS.

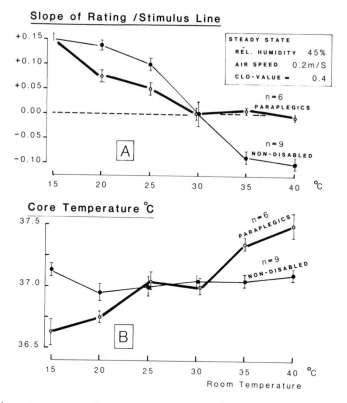

Fig. 2. Top: *Mean ± s.e. of rating/stimulus slope for a group of paraplegics and a control group of non-disabled subjects plotted against environmental temperature.*
    Bottom: *Mean ± s.e. of rectal temperture of a group of paraplegics and control group of non-disabled subjects plotted against environmental temperature. From: Attia & Engel (1983), courtesy of Churchill Livingstone.*

AMPUTEES AND DYSMELEES

    Directly after the operation and many years thereafter, amputees often complain of excessive sweating, even under resting conditions in an atmosphere of thermal comfort (Engel *et al.*, 1978). Long term observation of children with con- genital mal-development of the limbs (dysmelees) has shown that, in the early post delivery months, they sweat more and have comparatively higher core temperatures than non-disabled babies. In later years, when they are above the age of 12

years, dysmelees seem to be free of this unpleasant
experience (Engel, 1982). Children tended to adapt to the
new situation during a span of 12 years since their birth.
Adult amputees failed to adapt to the new situation and continued to suffer from excessive sweating and higher core
temperature and acquired no set-point shift which fits the
condition of loss of extremities.

FEVER

Fever is a primary disorder of temperature regulation and
a common clinical sign in many diseases. It is characterized
by an upward shift of the temperature regulatory set-point
caused by endogenous pyrogens (Stitt, 1981; Hensel, 1981;
Banet, 1979). As a result of this, the cold sensors increase
their firing rate, while the warm ones decrease it. The cold
defence mechanisms are thus activated and body temperature
rises until the signals from both types of sensors are again
of equal intensity (Banet, 1979). If the hypothalamus is
locally heated during onset of fever, the febrile responses
and the increase in body temperature can be suppressed
(Eisenman, 1972). By this method the load error caused by the
higher set temperature is diminished. It is important to
emphasize that during fever, the body retains its ability to
regulate its temperature (Stitt, 1981; Cabanac & Massonnet,
1974; Banet, 1979). In rats infected with Salmonella
enteritidis, Banet (1979, 1982) showed that the probability
of survival decreased parallel to the increase in body temperature, but it increased parallel to the increase in metabolic
rate. Banet (1982) concluded that the net effect of fever is
beneficial for the defences of the organism, if the metabolic
rate required to raise body temperature by 1°C is higher than
a critical minimum. Thus, the rise in body temperature during
fever is a chance unfavourable incident resulting from the
activation of cold defense mechanisms.

HEAT STROKE

Body core temperature levels in man of 42.0°C and above
were reported to be incompatible with life (Ferris *et al.*,
1938; Shibolet *et al.*, 1976). Such extreme body temperatures
were thought to denature enzymes, liquify membrane lipids,
damage mitochondria, affect coding of proteins and impair the

co-ordination of physiological processes, thus causing irreversible damage (Shibolet *et al.*, 1976). The concept of Critical Thermal Maximum (CTM) was coined by Cowles & Bogert (1944) and defined by Hutchison (1961) as the thermal point at which locomotory activity becomes disorganized and the animal loses its ability to escape from conditions that will promptly lead to its death. Bynum *et al.* (1978) attempted to redefine the concept of CTM in man as the combination of exposure duration and elevated body temperatures that result in subclinical or clinical heat injury. Hutchison (1980) corrected the misunderstanding introduced by Bynum *et al.* (1978) and stated that the CTM is an ecological maximum and not the physiological lethal temperature. Recently, however, heat stroke patients with rectal temperatures as high as 46.5°C (Khogali, 1983) and 47.0°C (Hart *et al.*, 1982) recovered, and thereby confirmed that the concept of CTM is valid. According to Hutchison (1980) the catastrophic clinical symptomatology which is suggestive of heat stroke in humans, fits the definition of the CTM. CTM or heat stroke in man is characterized by a total and partial failure of behavioural and autonomic temperature regulatory systems, respectively. Observations of heat stroke patients suggested that vasoconstriction, lack of sweating and in many cases shivering prevailed (Khogali & Al-Khawashki, 1981; Khogali, 1983).

Experimentally-induced heat stroke in sheep showed a similar tendency for cessation of open mouth panting, peripheral vasoconstriction and shivering (Khogali *et al.*, 1983). All heat stroke patients suffered from coma, presented with a hot dry skin and did not sweat at rectal temperatures ranging from 40.5 to 46.5°C. During emergency treatment on the Makkah Body Cooling Unit (MBCU), some patients went through a stage of shivering at high rectal temperatures of up to 41.0°C (Khogali, 1983). It is known that sweating threshold in man corresponds to an oesophageal temperature of 36.75°C (Hensel, 1981). Shivering and metabolic heat over-production is elicited by cold-receptors of the skin, but its magnitude is strictly controlled by central warm-inhibition. The inhibition becomes total when central temperature rises beyond the set-point at which the responses of sweating and vasodilatation begin, regardless even of the most intense skin cold receptor firing (Benzinger, 1979). The observed cutaneous vasoconstriction and the condition of lack of sweating as well as the presence of shivering in heat stroke patients suggest a sizable upward shift in temperature regulatory set-point. In sheep, open mouth panting was evident at rectal temperatures above 42.0°C, and continually increased in frequency proportional to the rise in core temperature accompanied by excessive thick salivation until a point was reached when a progressive marked drop in open mouth panting was noticed. At rectal temperatures of 44.0°C open mouth panting stopped concomitant with a drop in skin temperature on the lower leg and ear (Khogali *et al.*, 1983). Shortly after this state was reached, the sheep collapsed, went into convulsions and shivered. Heat stroke death occurred some 15 to 40 minutes after the onset of these symptoms in fleeced and defleeced sheep respectively. These results indicate that at extreme heat strain levels, the thresholds of panting, shivering and peripheral vasoconstriction are significantly shifted upwards. It is thus possible to argue that conditions typical of "hypothermia" prevailed; and a temperature regulatory set-point elevation is therefore assumed. The apparent upward shift of set-point in heat stroke is viewed as an *acute adjustment of thermal tolerance*. Evidence for this argument stems from the consideration of the concept of *heat hardening*, which was recently defined as a mechanism of acute adjustment of tolerance, requiring only minutes to a few hours (Maness & Hutchison, 1980; Hutchison & Maness, 1979). The question is whether the assumed shift serves the animal to escape death at core temperatures far beyond CTM. During sustained exposure to an excessive exogenous and/or endogenous heat load, when the body is fully wetted and the rate of heat accumulation is zero, the body can offset its load at a constant higher than normal core temperature and a state of thermal subjective discomfort. If the heat load increases beyond this limit, the rate of heat accumulation attains non-zero, positive values, body temperature rises steadily, maximum peripheral vasodilatation prevails and the subject suffers from extreme thermal discomfort. As more blood is drawn to the shell to meet increasing demands of temperature regulation, vital organs such as the heart and brain may suffer from partial deficiency in blood supply necessary to maintain its vital functions, part of which is to meet the afferent temperature regulatory requirements. It is not clear how these two conflicting demands could be met and which one overrides the other. At this critical stage the CNS seems to intervene. Experimental as well as clinical observations indicate that core requirements for blood override peripheral distribution necessary to maintain the temperature regulatory system intact. Hales (1983) argues that low-pressure baroreceptors could be responsible for dominance of circulatory over temperature control.

An upward shift in set-point is the only theoretical possibility which can explain the observed vasoconstriction, lack of sweating, cessation of panting and probably shivering during the heat stroke episode. Recovery and survival of heat stroke patients indicated that no irreversible damage was evident at core temperatures of up to 4.5°C above the CTM of 42.0°C (Khogali, 1983; Hart *et al.*, 1982). The projected thermal threshold (or CTM) of 42.0°C for humans is related to a *normal* reference temperature regulatory set-point of 37.1 ± 0.3°C (Attia *et al.*, 1980); with an elevation of the set-point it seems that the CTM is proportionally raised.

REFERENCES

Attia, M., Engel, P. & Hildebrandt, G. (1980). Thermal comfort during work: a function of time of day. *Int. Archs. Occup. Health 45*, 205-215.

Attia, M. & Engel, P. (1981). Thermal alliesthesial response in man is independent of skin location stimulated. *Physiol. Behav. 27*, 439-444.

Attia, M. & Engel, P. (1982). Thermal pleasantness sensation: an indicator of thermal stress. *Euro. J. Appl. Physiol. 50*, 55-70.

Attia, M. & Engel, P. (1983). Thermoregulatory set point in spinal cord injury patients. *Paraplegia* (in press).

Banet, M. (1979). Fever and survival in the rat. *Pflügers Arch. 381*, 35-38.

Banet, M. (1982). The survival value of fever. A temperature or a metabolic effect? *In* "Biological Adaptation" (G. Hildebrandt & H. Hensel, eds), pp. 210-211. G. Thieme Verlag Stuttgart, New York.

Benzinger, T.H. (1979). Physiological basis for thermal comfort. *In* "Indoor Climate" (P.O. Fanger & O. Valbjorn, eds), pp. 441-476. Danish Building Research Institute, Copenhagen.

Bynum, G.D., Pandolf, K.B., Schuette, W.H., Goldman, R.F., Lees, D.E., Whang-Peng, J., Atleinson, E.R. & Bull, J.M. (1978). Induced hyperthermia in sedated humans and the concept of critical thermal maximum. *Am. J. Physiol. 235*, R228-R236.

Cabanac, M. (1969). Plaisir ou deplaisir de la sensation thermique et homeothermal. *Physiol. Behav. 4*, 359-364.

Cabanac, M. (1972). Thermoregulatory behaviour. *In* "Essays
  on Temperature Regulation" (J. Bligh & R.E. Moore, eds),
  pp. 19-32. North Holland Publication Company, Amsterdam,
  London.
Cabanac, M. & Massonnet, B. (1974). Temperature regulation
  during fever: change of set point or change of gain? A
  tentative answer from a behavioural study in man.
  J. Physiol. (Lond.) 238, 561-568.
Cabanac, M., Hildebrandt, G., Massonnet, B. & Strempel, H.
  (1976). A study of the nycthemeral cycle of behavioural
  temperature regulation in man. *J. Physiol. (Lond.)
  257*, 275-291.
Cowles, R.B. & Bogert, C.M. (1944). A preliminary study of
  the thermal requirements of desert reptiles.
  Bull.Am. Museum Nat. Hist. 83, 265-296.
Cunningham, D.J. & Cabanac, M. (1971). Evidence from
  behavioural thermoregulatory responses of a shift in set
  point temperature related to the menstrual cycle.
  J. Physiol.(Paris) 63, 236-238.
Downey, J.A., Miller, J.M. & Darling, R.C. (1969). Thermo-
  regulatory responses to deep and superficial cooling in
  spinal man. *J. Appl. Physiol. 27*, 209-212.
Downey, J.A., Huckaba, C.E., Kelley, P.S., Tam, H.S.,
  Darling, R.C. & Cheh, H.Y. (1976). Sweating responses to
  central and peripheral heating in spinal man.
  J. Appl. Physiol. 40, 701-706.
Eisenman, J.S. (1972). Unit activity studies of thermo-
  responsive neurons. *In* "Essays on Temperature
  Regulation" (J. Bligh & R.E. Moore, eds), pp. 55-69.
  North Holland Publication Company, Amsterdam, London.
Engel, P., Rode, F.W., Schindler, W. & Attia, M. (1978).
  Anpassung der Thermoregulation an Gliedmassenverlust.
  S.F.B. 122. Philipps Universität Marburg/L. 5, 279-307.
  Band 5, 279-307.
Engel, P. (1982). Adaptation of thermoregulation to the loss
  of extremities. *In* "Biological Adaptation"
  (G. Hildebrant & H. Hensel, eds), pp. 198-208. G. Thieme
  Verlag Stuttgart, New York.
  G.E. (1938). Heat Stroke: clinical and chemical
  observations on 44 cases. *J. Clin. Invest. 17*, 249-262.
Guttmann, L. (1976). "Spinal Cord Injuries: Comprehensive
  Management and Research." Blackwell Scientific
  Publications, 2nd Ed., Melbourne.
Hales, J.R.S. (1983). Circulatory consequences of hyper-
  thermia: an animal model for studies of heat stroke. *In*
  "Heat Stroke and Temperature Regulation" (M. Khogali &
  J.R.S. Hales, eds), pp. 223-240. Academic Press, Sydney.

Hardy, J.D. (1972). Peripheral inputs to the central regulator for body temperature. *In* "Advances in Climatic Physiology" (S. Itoh, K. Ogata & H. Yoshimura, eds), pp. 3-21. Springer Verlag, Heidelberg, New York.

Hart, G.R., Anderson, R.J., Crumpler, C.P., Shulkin, A., Reed, G. & Knochel, J.P. (1982). Epidemic classical heat stroke: clinical characteristics and course of 28 patients. *Medicine 61*. 189-197.

Hensel, H., Brück, K. & Raths, P. (1973). Homeothermic organisms. *In* "Temperature and Life" (H. Precht, J. Christophersen, H. Hensel & W. Larcher, eds), pp. 503-733. Springer Verlag, Berlin, Heidelberg, New York.

Hensel, H. (1981). "Thermoreception and Temperature Regulation". Academic Press, London.

Hensel, H., Banet, M. & Schaefer, K. (1982). Central and peripheral components of thermal adaptation. *In* "Biological Adaptation" (G. Hildebrandt & H. Hensel, eds), pp. 134-139. G. Thieme Verlag, Stuttgart, New York.

Hildebrandt, G. (1982). The time structure of adaptive processes. *In* "Biological Adaptation" (G. Hildebrandt & H. Hensel, eds), pp. 24-39. G. Thieme Verlag, Stuttgart, New York.

Hutchison, V.H. (1961). Critical thermal maxima in salamanders. *Physiol. Zool. 34*, 92-125.

Hutchison, V.H. (1980). The concept of critical thermal maximum. *Am. J. Physiol. 237*, R367-R368.

Hutchison, V.H. & Maness, J.D. (1979). The role of behaviour in temperature acclimation and tolerance in ectotherms. *Amer. Zool. 19*, 367-384.

Khogali, M. (1983). Heat stroke: an overview. *In* "Heat Stroke and Temperature Regulation" (M. Khogali & J.R.S. Hales, eds), pp. 1-12. Academic Press, Sydney.

Khogali, M. & Al-Khawashki, M.I. (1981). Heat stroke during the Makkah Pilgrimage. *Saudi Medical Journal, 2*, 85-93.

Khogali, M., Elkhatib, G., Attia, M., Mustafa, M.K.Y., Gumaa, K., Nasralla, A. & Al-Adnani, M. (1983). Induced heat stroke: model in sheep. *In* "Heat Stroke and Temperature Regulation" (M. Khogali & J.R.S. Hales, eds), pp. 253-261. Academic Press, Sydney.

Maness, J.D. & Hutchison, V.H. (1980). Acute adjustment of thermal tolerance in vertebrate ectotherms following exposure to critical thermal maxima. *J. Thermal Biol. 5*, 225-233.

Paget, G.E. (1885). Case of remarkable risings and fallings of body temperature. *Lancet 2*, 4.

Pledger, H.G. (1962). Disorders of temperature regulation in acute traumatic tetraplegia. *J. Bone and Joint Surg.* *44-B*, 110–113.

Shibolet, S., Lancaster, M.C. & Danon, Y. (1976). Heat stroke: A review. *Aviat. Space Environ. Med.* *47*, 280–301.

Stitt, J.T. (1981). Neurophysiology of fever. *Fed. Proc.* *40*, 2835–2842.

7
Mechanism of Fever and Antipyresis

W. L. Veale, W. D. Ruwe and K. E. Cooper

Department of Medical Physiology
University of Calgary
Calgary, Alberta, Canada

Fever is the pathological state in which body temperature is elevated and regulated. Exogenous pyrogens originate outside the host and result in the production of endogenous pyrogen (EP). EP has been shown to be produced by many different reticuloendothelial system cell types and it is thought that EP is carried in the circulation to the brain, where it acts on neurons in the anterior hypothalamic/preoptic area (AH/POA) to produce fever.

The AH/POA has been shown to be the site in the brain sensitive to pyrogenic effects of both exogenous and endogenous pyrogens. Fever is known to raise the set-point around which body temperature is regulated and may be distinguished from other hyperthermias such as that resulting from exercise. In exercise the new temperature is not actively defended.

Prostaglandins of the E series (PGE) have a potent pyrogenic effect when microinjected into the brain. There is, however, evidence that PGE may not be involved in all types of fevers.

Another putative neurotransmitter specifically involved in the control of fever is vasopressin. This peptide has been shown to act as an endogenous antipyretic in the brain of both the sheep and rat and, as such, may form a negative feedback loop in the control of fever.

HEAT STROKE AND
TEMPERATURE REGULATION
ISBN 0 12 406180 X

BASIC DEFINITIONS

A. *Fever*

Fever is recognized as a pathological process by which a regulated increase in core temperature occurs during infection. The mechanisms underlying the pathogenesis of fever have been examined extensively and Figure 1 illustrates a widely accepted schema for the development of fever. (1-9). The febrile response can be initiated by any one of a number of activators.

B. *Endotoxins*

Endotoxins, gram-positive bacteria and viruses are among the most frequently encountered exogenous pyrogens. Endotoxin, derived from the cell walls of gram-negative bacteria, is a large lipopolysaccharide with a molecular weight of one to two million (10-12). A basal core region of polysaccharides, an O-specific side chain and a lipid

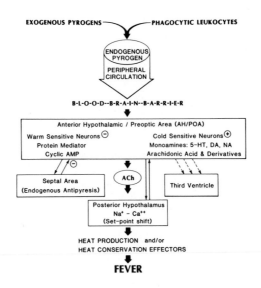

Fig. 1.   Scheme for fever pathogenesis.

moiety constitute the three primary components of the endotoxin. Currently, it is believed that the pyrogenic activity of the molecule is derived from the lipidmoiety or lipid A region (4,7,13,14). A very potent activator of fever, systemic administration of endotoxin in doses as small as 0.0001 micrograms per kilogram produces a marked elevation in the core temperature of the rabbit (15-17).

C. *Endogenous Pyrogen*

The production and release of another pyrogenic substance, endogenous pyrogen, occurs as exogenous bacteria are destroyed and absorbed by various phagocytic cells within the host organism (7). Endogenous pyrogen is a heat-labile protein with a molecular weight of approximately 15,000 daltons which may be present in the peripheral circulation as a trimer. Although an organism rapidly develops tolerance to repeated injections of endotoxin, little or no tolerance results from repeated administrations of endogenous pyrogens. Initial experimental evidence suggested that only polymorphonuclear granulocytes produced endogenous pyrogen (18), however, it is now known that many types of immunologically active cells of the reticuloendothelial system are capable of producing this substance (2). Endogenous pyrogens may be produced by neutrophils (19,20), monocytes (21), eosinophils (22), splenic sinusoidal cells (23), lung alveolar macrophages (23), liver macrophages (24,25), peritoneal lining cells (23) and other phagocytic cells derived from bone marrow precursors.

There is very little evidence to suggest that exogenous pyrogens can enter the central nervous system from the periphery. It has been speculated that the blood-brain barrier is impervious to penetration by endotoxin owing to the extremely large size of most exogenous pyrogens (4). Unlike endotoxin, endogenous pyrogen may cross the blood-brain barrier (7). During the late fifties, King and Wood (26) provided the first direct evidence that the brain is involved in the pathogenesis of fever. In contrasting the effects of infusing an endogenous pyrogen into the carotid arteries or into the venous circulation, these investigators observed a fever of much shorter latency and of far greater magnitude following an intracarotid injection of the pyrogen (4). Through subsequent investigation, the anterior hypothalamic, preoptic area or AH/POA has been shown to be the most sensitive area of the brain to pyrogenic material (27-32).

THERMOSENSITIVITY

The spontaneous activity of both warm-sensitive (increase firing rate as temperature increases) and cool-sensitive (increase firing rate as temperature decreases) neurons in the AH/POA has been demonstrated to be affected by the administration of pyrogenic material.

Following administration of exogenous pyrogens (33–37) or endogenous pyrogen (36,38) the electrical activity of the thermosensitive neurons in the AH/POA is altered. In almost 100% of the warm-sensitive neurons observed, administration of pyrogen inhibits the firing rate, whereas, there is a pyrogen-induced facilitation of the firing rates in over 90% of the cool-sensitive neurons (39).

CHEMOSENSITIVITY

A. *Monoamines*

Monoaminergic involvement in the pathogenesis of fever has been the subject of much controversy. Only with extremely large doses is the depleting agent reserpine able to alter the typical febrile response to administration of either endotoxin or endogenous pyrogen (31,40–43). Recently, investigators have utilized more selective pharmacological agents than reserpine, which depletes central stores of catecholamines and indoleamines. Even these have failed to clearly elucidate the role that each amine may play in the genesis of fever. Selective depletion of catecholamine stores with alpha-methyl-p-tryrosine (α-MPT) or alpha-methyl-m-tyrosine (α-MMT) does not markedly affect the febrile response (40,41). However, some evidence does exist to suggest that α-MPT may reduce the magnitude of an endotoxin-induced fever (41). Similarly, a pyrogen-induced fever may be diminished by administration of 6-hydroxydopamine (42,44).

B. *Prostaglandins*

The role of prostaglandins in the development of the febrile response has received much attention in recent years. In 1971, Milton and Wendlandt (45) injected an E-series prostaglandin into the cerebral ventricle of the cat and rabbit and observed a profound hyperthermia with a short latency of onset. These findings later were replicated in a number of different species including the sheep and the rat (46-48). One of the many naturally occurring constituents of the hypothalamus (49,50), prostaglandin E_2 activates heat production mechanisms and inhibits heat-loss processes in a manner very similar to that of exogenous pyrogen or endogenous pyrogen following injection into local AH/POA loci (51,52). Prostaglandin-like material has been detected with bioassay systems in the cerebrospinal fluid of an animal following intravenous or intraventricular injections of an exogenous pyrogen (53,54). Moreover, employing more sensitive radioimmunoassay techniques, prostaglandin levels nearly twice the normal concentration have been measured in the cerebrospinal fluid of an animal infected with a bacterium or a virus (55). Additional evidence which favours the prostaglandins as a critical mediator of fever is derived from investigations in which antipyretics are administered. Acetylsalicylic acid, 4-acetamidophenol and indomethacin which apparently attenuate fever by acting on the enzyme systems concerned with the synthesis of prostaglandins (56), diminish or prevent the increase in the level of prostaglandins in the cerebrospinal fluid concomitantly with the reduction in body temperature (54).

In 1974 Baird and her colleagues (57) demonstrated that, although an intravenous infusion of an exogenous pyrogen evoked a fever, prostaglandins injected into the lateral cerebral ventricle of *Tachyglossus Sp.*, echidna, decreased the body temperature of this primitive monotreme mammal. This report provided the first evidence that prostaglandins might not be the sole neurochemical mediary underlying a pyrogen-induced fever. It was later discovered that newborn lambs respond in a manner not unlike that of the echidna (58). Peripheral or central administration of exogenous pyrogen elicited a normal febrile response, whereas, following an intrahypothalamic injection of a prostaglandin, fever did not occur.

Veale and Cooper (59) found that an intravenous injection of endogenous pyrogen could still elicit a fever in the rabbit without an intact AH/POA. In this same AH/POA-lesioned animal, however, an intrahypothalamic microinjection of a prostaglandin did not cause an increase in body temperature. Thus, additional evidence was provided for the dissociation between a fever caused by bacterial infection and the rise in body temperature normally evoked by injections of prostaglandins.

Still more convincing evidence that prostaglandins are not the essential mediator of a pyrogen-induced fever has come from studies utilizing single unit recording techniques. If it is assumed that pyrogens exert their action on neurons within the AH/POA via the prostaglandins, then one would expect that both pyrogens and prostaglandins would affect the firing rates of these cells in a similar fashion. Yet, whereas application of an exogenous pyrogen causes an increase in the firing rate of a cool-sensitive neuron, prostaglandins appear to mildly enhance the firing rate of the warm-sensitive neuron (60). However, it is not yet possible to elaborate a definitive statement with regard to the implications of those studies as other investigators have reported conflicting results (61).

Cranston and his colleagues (62) demonstrated that endogenous pyrogens infused into the vein of a rabbit produce an increase in body temperature which is concurrent with an increase in circulating levels of prostaglandins within the cerebrospinal fluid. However, following systemic pretreatment with a very small dose of an antipyretic, the animal becomes febrile, but there are no measurable differences in the concentrations of prostaglandins in the cerebrospinal fluid (62). These same investigators obtained similar results with either one of two potent prostaglandin antagonists. These studies suggest that a close derivative of prostaglandin, rather than the substance itself, may be involved in the febrile response.

C. Protein Mediators

Siegert, Philipp-Dormston and their colleagues (63,64) first reported that cycloheximide, a potent inhibitor of protein synthesis, interferes with the development of a pyrogen-induced fever. These initial observations raised the possibility that an unknown protein factor could mediate the intense hyperthermia which characterizes the febrile response. A number of subsequent investigations have indicated that cycloheximide not only prevents the development of a fever produced by an exogenous pyrogen (65-67), but that the typical febrile response to an endogenous pyrogen can be prevented by similar administration of this protein synthesis inhibitor (67-70). Owing to cycloheximide's nonspecific interference with the body's metabolic processes (65,71,72) and the generalized toxicity of cycloheximide (73-78), the findings using this compound have been called into question. For these reasons, anisomycin, another potent inhibitor of protein synthesis has been utilized to determine the role of protein synthesis in the pathogenesis of fever. This antibiotic is characterized by its structural specificity, short-lived action, reversibility and minimal toxicity (79-84).

When anisomycin is administered either peripherally or centrally, into the ventricular spaces or hypothalamic loci, the febrile response normally evoked by a similar administration of exogenous pyrogen (84-86) or endogenous pyrogen (85,87,88) is attenuated or prevented. Moreover, this inhibitor of protein synthesis does not appear to alter the normal thermoregulatory responses to the prostaglandins (84,85) or to the monoamines (84,85). Indeed, the thermoregulatory system of the animal appears to remain completely intact following administration of anisomycin (85,87). Most recently, it has been demonstrated that a number of protein synthesis inhibitors attenuate or prevent a pyrogen-induced fever (89), whereas, comparable compounds such as analogues of anisomcycin, (GS-5545 and GS-5546) which lack protein synthesis inhibiting actions do not affect the febrile response (90). Overall, these experiments have provided solid evidence that a factor which is dependent upon the synthesis of new protein is ultimately necessary for the elaboration of a pyrogen-induced fever.

SET-POINT MECHANISMS

In 1970, Feldberg and his colleagues (91) demonstrated that an increase in the body temperature of the cat, evoked by an intraventricular infusion of sodium ions, could be antagonized by a similar administration of calcium ions (91). This finding gave rise to the proposition that a dynamic balance between these two cations could serve as a basis for the set-point mechanism. Subsequent investigations revealed that the neuroanatomical locus of this action was not in the AH/POA as one might have expected, but rather, the cations exerted their profound effects on body temperature through an action on neurons located within the posterior hypothalamus (92). Only at very circumscribed loci in the posterior hypothalamus of an animal does an infusion of excess sodium ions evoke a marked increase in body temperature. Shivering, vasoconstriction and piloerection occur concomitantly with the rise in temperature. When these same sites are perfused, using push-pull cannulae, with a solution containing excess calcium ions, a sharp fall in body temperature and vasodilatation occur.

In contrast to the effects of most putative neurotransmitters on temperature regulation and fever, there appears to be very little interspecies variation in the temperature response to the experimental manipulation of sodium and calcium concentrations within the posterior hypothalamus (93,94). Calcium hypothermia and sodium hyperthemia can be evoked by intrahypothalamic perfusions in the caudal diencephalon of the cat (92,95,96), rabbit (94,97,98), rat (99-101) hamster (102), ground squirrel (103,104), dog (105), rhesus monkey (106), pig-tailed monkey (107), and squirrel monkey (Jones, Veale and Cooper, unpublished observations).

Direct evidence for the Na^+ - Ca^{++} balance in the set-point function is derived from the experimental investigation of fever. Myers and Tytell (108) labelled the endogenous stores of calcium and sodium with $^{45}Ca^{++}$ or $^{22}Na^+$ and observed the subsequent fluctuations in these radiolabelled cations. When an exogenous pyrogen was administered systemically or intracerebrally, a reciprocal shift in the ratio of the endogenous cations in the diencephalon was evoked as the fever developed. The increase in the set-point around which body temperature is regulated during fever can be distinguished from other hyperthermias such as that resulting from exercise, a state in which the new temperature is not actively defended against alterations in ambient temperature (109-112).

SURVIVAL VALUE OF FEVER

Does fever increase the likelihood of survival during infection? Is fever, in fact, a host response which beneficially defends against infection? Although not new questions, only within the last decade have investigators begun a major thrust toward answering them. Evidence in support of the hypothesis that fever does benefit an organism has been obtained for fish (113,114), crayfish (115), amphibians (116-118), reptiles (119-121), ferrets (122) and rabbits (123,124). Fever could be beneficial to an organism in one of two ways. First, the elevation in body temperature could impede the growth rate of an infectious microorganism or, perhaps, directly kill it. Second, fever could have an indirect influence on the microorganism by triggering a specific biochemical, cellular or humoral component of the body which, in turn, destroys the infectious agent. The high body temperature could increase lysosome breakdown, increase the production of interferon, alter some aspect of leukocyte function or activate some other as yet unknown mechanism or mechanisms. Increases in the mobility of leukocytes, the amount of phagocytic activity, in the bactericidal activity of leukocytes intracellularly and/or in the proliferation and transformation of lymphocytes are known to occur as a result of temperatures comparable to those observed during a moderate fever in humans (125).

EXOGENOUS ANTIPYRESIS

"Substances which reduce the temperature in febrile and similar states but not in normal conditions, unless the dosage be excessive, are termed antipyretics". This description of antipyretics first used by Barbour in 1921 (126) is still accepted today as an accurate and adequate description of both the actions of and clinical uses for antipyretics. As illustrated in Figure 2, there are a

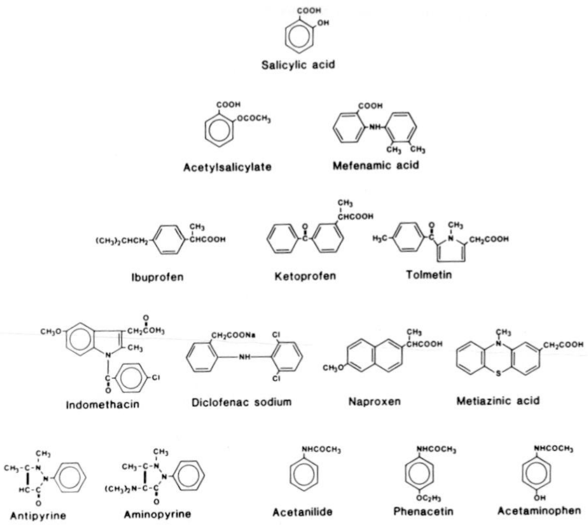

Fig. 2. Structures of some antipyretics.

number of exogenous antipyretic substances which have proven useful in reducing the body temperature of a febrile animal. There is, however, no clear indication as to the means by which the febrile state is attenuated or abolished. Antipyresis could be effectively initiated at any one of several stages in the pathogenetic processes which occur during infection. Thus, an antipyretic could: 1) interact with exogenous pyrogen; 2) interfere with the synthesis and/or release of EP by cells of the reticuloendothelial system; 3) biochemically inactivate the EP molecule in the blood; 4) accelerate the elimination of EP from the circulation; 5) prevent passage of EP from the blood into the brain; 6) interfere with the central action of EP on neurons within the AH/POA or some other diencephalic structure; or 7) affect specific thermoregulatory effector organs or pathways of the organism (127,128). Current evidence indicates that the antipyretics act principally at the level of the AH/POA or some other brainstem structure through an action on the neuronal activity in these areas. Antipyresis might be achieved by: 1) interrupting the biosynthesis, release or synaptic action(s) of the biogenic amines, acetylcholine, cyclic nucleotides, prostaglandins or neuropeptides (e.g. 56); 2) antagonizing a specific EP receptor (e.g. 129); 3) altering indirectly the set-point mechanism, Na^+-Ca^{++} ratio, (e.g. 130); 4) inducing the release of an endogenous antipyretic (e.g. 131–133); or 5) activating an integrated response system involving more than one of these mechanisms.

ENDOGENOUS ANTIPYRESIS

Another putative neurotransmitter involved in the modulation of fever is arginine vasopressin or AVP. This neuroactive peptide has been shown to act as an endogenous antipyretic in the mammalian brain and, as such, may form a negative feedback loop in the control of fever. The postulated existence of this endogenous antipyretic is primarily the result of a very extensive investigation conducted in the laboratory of Veale and Cooper. Investigators in this laboratory initially observed that the animal approaching parturition does not develop a fever to exogenous pyrogen nor to EP. Thus, the febrile response is markedly diminished or completely absent from nearly four days prepartum until several hours after delivery (131,132). Similarly, there is an absence of the febrile response to pyrogen in the neonate despite its ability to adequately thermoregulate (131,132).

During that period of time in which the febrile response is markedly reduced, the plasma concentration of arginine vasopressin is at its highest levels (134). As term approaches AVP levels increase, then subsequently return to basal levels shortly after delivery. This hormonal fluctuation is highly correlated with the period of diminished febrile response.

Zeisberger and his colleagues (135,136) have reported that the periparturient guinea pig is incapable of generating a fever several days before term and several hours postpartum. These same investigators also demonstrated an increase in the amount of AVP-reactive material in the neuronal elements of the septo-hypothalamic system in the pregnant animal (137,138). These changes were most prominent during the prepartum period, extending through term until one day postpartum, at which time the basal levels of AVP-staining material is observed in these nuclear regions.

Kasting and his colleagues subsequently have demonstrated that AVP may function as an endogenous antipyretic in the non-pregnant animal (133,139). Although the perfusion of AVP into sites in the brain of an afebrile animal is without effect on core temperature, when this nonapeptide is infused into these same sites, a pyrogen-induced fever is markedly suppressed. This antagonism of the febrile response can be effected with doses of less than 10.0 micrograms.

The sites of maximum sensitivity to this neuropeptide appear to reside within the septal area, only 2 to 3 mm rostral to the anterior commissure (133). Strangely enough, AVP does not affect the core temperature of an animal following perfusion within the classical thermoregulatory areas of the diencephalon, including the AH/POA and the posterior hypothalamus.

Although AVP reduces the absolute magnitude of a pyrogen-induced fever in a dose-dependent manner, the comparable infusion of oxytocin, somatostatin, angiotensin II or substance P does not affect the fever. A structurally similar peptide which elicits many of the same peripheral effects of AVP, arginine vasotocin, only partially antagonizes the fever (140). Bombesin, a potent thermolytic agent, also reduces the body temperature when infused into these central septal sites. However, since this peptide is equally effective in reducing the core temperature of an afebrile animal, it is likely that its action is non-specific in nature.

The central release of AVP is altered within the AVP-sensitive sites in the septal region during the development of fever (133). Analysis of push-pull perfusates by radioimmunoassay reveals decreasing levels of AVP are released into the extracellular fluid of this region as body temperature increases. In marked contrast, as body temperature decreases there is an upward shift in the amount of AVP released into the perfusates obtained from these septal sites.

The typical febrile response of an animal can be clearly enhanced by infusion of either a specific antiserum to AVP or an AVP analogue (desamino-dicarba-AVP) into the septal area (8). These findings suggest that an intact receptor population as well as the proper neurochemical function of AVP are requisite to the suppressive action of AVP during fever.

The antipyretic action of AVP has been reported recently for several strains of the rat (141) and the rabbit (Ruwe, Veale, Nagata & Disturnal, unpublished observations).

FEVER AND CONVULSIVE DISORDERS

In the rat, AVP also has been demonstrated to evoke seizures or seizure-like activity (142). Behavioural depression, including immobility and staring are observed upon initial administration of AVP, followed by seizures with myoclonic-myotonic convulsive episodes upon subsequent administration of the peptide. In addition, animals which lack AVP synthesizing capabilities, the Brattleboro (DI) rat, or those in which AVP stores have been neutralized with a specific anti-AVP antiserum exhibit a much higher threshold in the core temperature at which heat-induced seizures occur (142,143). These preliminary findings suggest that AVP may play a role in the etiology of febrile convulsions.

Acknowledgement

Thanks to G. Olmstead for typing this manuscript.

REFERENCES

1. Atkins, E. and Bodel, P., *N. England J. Med. 286*, 27-34 (1972).
2. Atkins, E. and Bodel, P., *in* "The Inflammatory Process. Volume III" (B.W. Zweifach, L. Grant and R.T. McCluskey, eds.), pp. 467-514. Academic Press, New York, (1974).
3. Atkins, E. and Bodel, P., *Federation Proc. 38*, 57-63 (1979).
4. Dinarello, C.A. and Wolff, S.M., *in* "Brain Dysfunction in Infantile Febrile Convulsions" (M.A.B. Brazier and R. Coceani, eds.), pp. 117-128. Raven Press, New York, (1976).
5. Dinarello, C.A. and Wolff, S.M., *N. England J. Med. 298*, 607-612 (1978).
6. Dinarello, C.A., *Fed. Proc. 38*, 52-56 (1979).
7. Kluger, M.J. Fever: Its Biology, Evolution and Function. Princeton, New Jersey: Princeton University Press, (1979).
8. Kasting, N.W., *Ph.D. Thesis*, University of Calgary, Calgary, Canada, 1980.
9. Bernheim, H.A., Block, L.H. and Atkins, E., *Ann. Intern. Med. 91*, 261-270 (1979).
10. Luderitz, O., Staub, A.M. and Westphal, O., *Bacteriol. Rev. 30*, 192-255 (1966).
11. Nowotny, A., *Bact. Rev. 33*, 72-98 (1969).
12. Bennett, I.L., Jr. and Beeson, P.B., *Medicine 29*, 365-400 (1950).
13. Luderitz, O., Galanos, C., Lehmann, V., Murnimen, M., Rietschel, E.T., Rosenfelder, G., Simon, M. and Westphal, O., *J. Infec. Dis. 128(Suppl)*, S17-S29 (1973).
14. Milton, A.S., *J. Pharm. Pharmac. 28*, 393-399 (1976).
15. Landy, M. and Johnson, A.G., *Proc. Soc. Exp. Biol. Med. 90*, 57-62 (1955).
16. Snell, E.S. and Atkins, E., *in* "The Biological Basis of Medicine, Vol. 2" (E.E. Bittar and N. Bittar, eds.), pp. 397-419. Academic Press, New York, (1968).
17. van Miert, A.S.J.P.A.M. and Frens, J., *Zbl. Vet. Med. 15*, 532-543 (1968).
18. Herion, J.C., Walker, R.I. and Palmer, J.G., *J. Expt. Med. 113*, 1115-1125 (1961).
19. Bennett, I.L., Jr. and Beeson, P.B., *J. Exp. Med. 98*, 477-492 (1953).

20. Bennett, I.L., Jr. and Beeson, P.B., *J. Exp. Med. 98*, 493-508 (1953).
21. Bodel, P. and Atkins, E., *Proc. Sec. Exp. Biol. Med. 121*, 943-946 (1966).
22. Mickenberg, I.D., Root, R.K. and Wolff, S.M., *Blood 39*, 67-80 (1972).
23. Atkins, E., Bodel, P. and Francis, L., *J. Exp. Med. 126*, 357-383 (1967).
24. Dinarello, C.A., Bodel, P. and Atkins, E., *Trans Assoc. Am. Physicians 81*, 334-343 (1968).
25. Haesler, F., Bodel, P.T. and Atkins, E., *J. Reticeuloendothel Soc. 22*, 569-581 (1977).
26. King, M.K. and Wood, W.B. Jr., *J. Exp. Med. 107*, 291-303 (1958).
27. Villablanca, J. and Myers, R.D., *Arch. Biol. Med. Exp. 1*, 102 (1964).
28. Villablanca, J. and Myers, R.D., *Am. J. Physiol. 208*, 703-707 (1965).
29. Jackson, D.L., *J. Neurophysiol. 30*, 586-602 (1967).
30. Repin, I.S. and Kratskin, I.L., *Fiziol. Zh. SSSR 53*, 1206-1211 (1967).
31. Cooper, K.E., Cranston, W.I. and Honour, A.J., *J. Physiol. 191*, 325-337 (1967).
32. Myers, R.D., Rudy, T.A. and Yaksh, T.L., *J. Physiol. 243*, 167-193 (1974).
33. Cabanac, M., Stolwijk, J.A.J. and Hardy, J.D., *J. Appl. Physiol. 24*, 645-652 (1968).
34. Wit, A. and Wang, S.C., *Am. J. Physiol. 215*, 1160-1169 (1968).
35. Eisenman, J.S., *Am. J. Physiol. 216*, 330-334 (1969).
36. Belyavskii, E.M. and Abramova, E.L., *Bull. Eksp. Biol. Med. 80*, 17-20 (1975).
37. Nakayama, T. and Hori, T., *J. Appl. Physiol. 34*, 351-355 (1973).
38. Schoener, E.P. and Wang, S.C., *Am. J. Physiol. 229*, 185-190 (1975).
39. Eisenman, J.S., *in* "Handbook of Experimental Pharmacology: Vol. 60 Pyretics and Antipyretics". (A.S. Milton, ed.), pp. 187-217. Springer-Verlag: Berlin, Heidelberg, New York, (1982).
40. Des Prez, R., Helman, R. and Oates, J.A., *Proc. Soc. Exp. Biol. Med. 122*, 746-749 (1966).
41. Teddy, P.J., *in* "Pyrogens and Fever. (G.E.W. Wolstenholme and J. Birch, eds.), pp. 124-127. Edinburgh: Churchill Livingstone, (1971).

42. Woolf, C.J., Willies, G.H., Laburn, H. and Rosendorff, C., *Neuropharmacology 14*, 397-403 (1975).

43. Lipton, J.M. and Trzcinka, G.P., *Am. J. Physiol. 231*, 1638-1648 (1976).

44. Borsook, D., Laburn, H.P., Rosendorff, C., Willies, C.H. and Woolf, C.J., *J. Physiol. 266*, 423-433 (1977).

45. Milton, A.S. and Wendlandt, S., *J. Physiol. 218*, 325-336 (1971).

46. Hales, J.R.S., Bennett, J.W., Baird, J.A. and Fawcett, A.A., *Pflügers Arch. 339*, 125-133 (1973).

47. Bligh, J. and Milton, A.S., *J. Physiol. 229*, 30-31P (1973).

48. Feldberg, W. and Saxena, P.N., *J. Physiol. 249*, 601-615 (1975).

49. Ambache, N., Brummer, H.C., Rose, J.G. and Whiting, J., *J. Physiol. 185*, 77-78P (1966).

50. Holmes, S.W. and Horton, E.W., *J. Physiol. 195*, 731-741 (1968).

51. Feldberg, W. and Saxena, P.N., *J. Physiol. 217*, 547-556)1971).

52. Stitt, J.T., *J. Physiol. 232*, 163-179 (1973).

53. Feldberg, W. and Gupta, K.P., *J. Physiol. 228*, 41-53 (1973).

54. Feldberg, W., Gupta, K.P., Milton, A.S. and Wendlandt, S., *J. Physiol. 234*, 279-303 (1973).

55. Philipp-Dormston, W.K. and Siegert, R., *Med. Microbiol. Immunol. 159*, 279-284 (1974).

56. Vane, J.R., *Nature New Biology 231*, 232-235 (1971).

57. Baird, J.A., Hales, J.R.S. and Lang, W.J., *J. Physiol. 236*, 539-548 (1974).

58. Pittman, Q.J., Veale, W.L. and Cooper, K.E., *Neuropharmacology 16*, 743-749 (1977).

59. Veale, W.L. and Cooper, K.E., *in* "Temperature Regulation and Drug Action." (P. Lomax, E. Schönbaum and J. Jacob, eds.), pp. 218-226. Basel: Karger, (1975).

60. Stitt, J.T. and Hardy, J.D., *Am. J. Physiol. 229*, 240-245 (1975).

61. Ford, D.M., *J. Physiol. 242*, 142-143P (1974).

62. Cranston, W.I., Hellon, R.F. and Mitchell, D., *J. Physiol. 253*, 583-592 (1975).

63. Philipp-Dormston, W.K. and Siegert, R., *Med. Microbiol. Immunol. 161*, 11-13 (1975).

64. Philipp-Dormston, W.K., *Zbl. Bakt. Hyg. 235*, 42-47 (1976).

65. Stitt, J.J., *in* "Thermoregulatory Mechanisms and Their Therapeutic Implications". (B. Cox, P. Lomax, A.S. Milton, E. Schönbaum, eds.), pp. 120-125. Karger: Basel, New York, London, (1980).
66. Milton, A.S. and Sawhney, V.K., *Br.J. Pharmacol. 70*, 97P (1980).
67. Milton, A.S. and Sawhney, V.K., *in* "Advances in physiological Sciences. Contributions to Thermal Physiology" (Z. Szelényi and and M. Székely, eds.), pp. 165-167. Akadémiai Kiadó, Budapest, (1980).
68. Cranston, W.I., Hellon, R.F., Luff, R.H. and Townsend, Y., *J. Physiol.*, 44P (1980).

69. Hellon, R.D., Cranston, W.I., Townsend, Y., Mitchell, D., Dawson, N.J. and Duff, G.W., *in* "Fever" (J.M. Lipton, ed.), pp. 159-164, Raven Press, New York, (1980).
70. Milton, A.S. and Todd, D., *Br. J. Pharmacol. 72*, 543P (1981).
71. Barney, C.C., Katovich, J.J. and Fregly, M.J., *Brain Res. Bull. 4*, 355-358 (1979).
72. Barney, C.C., Fregly, M.J., Katovich, M.J. and Tyler, P.E., *in* "Fever" (J.M. Lipton, ed.), pp. 111-122. Raven Press, New York, (1980).
73. Young, C.W., Robinson, P.F. and Sacktor, B., *Biochem. Pharmacol. 12*, 855-865 (1963).
74. Garren, L.D., Ney, R.L. and Davis, W.W., *Proc. Natl. Acad. Sci. USA 43*, 1443-1450 (1965).
75. Grahame-Smith, D.G., *J. Neurochem. 19*, 2409-2422 (1972).
76. Young, C.W. and Dowling, M.D. Jr., *Cancer Res. 35*, 1218-1224 (1975).
77. Ch'ih, J.J., Olszyna, D.M. and Devlin, T.M., *Biochem. Pharmacol. 25*, 2407-2408 (1976).
78. Garcia-Sainz, J.A., Pina, E. and De Sanchez, V.C., *Biochem. Pharmacol. 27*, 1577-1579 (1978).
79. Grollman, A.P. and Walsh, M., *J. Biol. Chem. 242*, 3226-3233 (1967).
80. Pestka, S., *Ann. Rev. Microbiol. 25*, 487-562 (1971).
81. Schwartz, J., Castellucci, V. and Kandel, E., *J. Neurophysiol. 34*, 939-953 (1971).
82. Flood, J.F., Rosenweig, M.R., Bennett, E.L. and Orme, A.E., *Behav. Biol. 10*, 147-160 (1974).
83. Flood, J.F., Bennett, E.L., Orme, A.E., Rosenweig, M.R. and Jarvik, M.E., *Science 199*, 324-326 (1978).

84. Ruwe, W.D. and Myers, R.D., *Brain Res. Bull. 4*, 741-745 (1979).
85. Ruwe, W.D. and Myers, R.D., *Brain Res. Bull. 5* 735-743 (1980).
86. Milton, A.S. and Sawhney, V.K., *Br. J. Pharmacol. 74*, 786P (1981).
87. Cranston, W.I., Hellon, R.F. and Townsend, Y., *J. Physiol. 305*, 337-344 (1980).
88. Cranston, W.I., Hellon, R.F. and Townsend, Y., *J. Physiol. 322*, 441-445 (1982).
89. Cannon, M., Cranston, W.I., Hellon, R.F. and Townsend, Y., *J. Physiol. 322*, 447-455 (1982).
90. Milton, A.S and Sawhney, V.K., *in* "Pyretics and Antipyretics. Handbook of Experimental Pharmacology, Vol. 60", (A.S. Milton, ed.), pp. 305-316. Springer-Verlag: Berlin, Heidelberg, New York, (1982).
91. Feldberg, W., Myers, R.D. and Veale, W.L., *J. Physiol. 207*, 403-416 (1970).
92. Myers, R.D. and Veale, W.L., *Science 170*, 95-97 (1970).
93. Veale, W.L. and Cooper, K.E. *in* "The Pharmacology of Thermoregulation". (E. Schönbaum and P. Lomax, eds.), pp. 289-301. Karger, Basel, (1973).
94. Jones, D.L., Veale, W.L. and Cooper, K.E., *Can. J. Physiol. Pharmacol. 56*, 571-576 (1978).
95. Myers, R.D. and Veale, W.L., *J. Physiol. 212*, 411-430 (1971).
96. Myers, R.D., Simpson, C.W., Higgins, D., Nattermann, R.A., Rice, J.C., Redgrave, P. and Metcalf, G., *Brain Res. Bull. 1*, 301-327 (1976).
97. Feldberg, W. and Saxena, P., *J. Physiol. 248*, 247-271 (1970).
98. Veale, W.L., Benson, M.J. and Malkinson, T., *Brain Res. Bull. 2*, 67-69 (1977).
99. Myers, R.D. and Brophy, P.D., *Neuropharmacology 11*, 351-361 (1972).
100. Gisolfi, C.V., Wilson, N.C., Myers, R.D. and Phillips, M.I., *Brain Res. 101*, 160-164 (1976).
101. Myers, R.D., Melchior, C.L. and Gisolfi, C.V., *Brain Res. Bull. 1*, 33-45 (1976).
102. Myers, R.D. and Buckman, J.D., *Am. J. Physiol. 223*, 1313-1318 (1972).
103. Hanegan, J.L. and Williams, B.A., *Science 181*, 663-664 (1973).
104. Hanegan, J.L. and Williams, B.A., *Comp. Biochem. Physiol. 50*, 247-252 (1975).

105. Sadowski, B. and Szczepanska-Sadowska, E., *Pflügers Arch. 352*, 1-68 (1974).

106. Myers, R.D., Veale, W.L. and Yaksh, T.L., *J. Physiol. (Lond.) 217*, 381-392 (1971).

107. Gisolfi, C.V., Myers, R.D. and Mora, F., *Fed. Proc. 36*, 1264 (1977).

108. Myers, R.D. and Tytell, M., *Science 178*, 765-767 (1972).

109. Nadel, E.R., *in* "Problems with Temperature Regulation During Exercise". (E.R. Nadel, ed.), pp. 1-10. Academic Press. New York, (1977).

110. Stitt, J.T., *Fed. Proc. 38*, 39-43 (1979).

111. Savard, G.K., Cooper, K.E. and Veale, W.L., *Can. Fed. Biol. Soc. 24*, 136 (1981).

112. Savard, G.K., Cooper, K.E. and Veale, W.L., *Experientia* (in press) 1983.

113. Reynolds, W.W., Casterlin, M.E., Covert, J.B., *Nature 259*, 41-42 (1976).

114. Covert, J.B., Reynolds, W.W., *Nature 267*, 43-45 (1977).

115. Casterlin, M.E., Reynolds, W.W., *Hydrobiologia 56*, 99-101 (1977).

116. Casterlin, M.E. Reynolds, W.W., *Life Sci. 20*, 593-596 (1977).

117. Kluger, M.J., *Thermobiology 2*, 79-81 (1977).

118. Myhre, K., Cabanac, M., Myhre, G., *Acta. Physiol. Scand. 101*, 219-229 (1977).

119. Kluger, J.J., Ringler, D.H., Anver, M.R., *Science 188*, 166-168 (1975).

120. Bernheim, H.A., Kluger, M.J., *Am. J. Physiol. 231*, 298-303 (1976).

121. Bernheim, H.A., Kluger, M.J., *Science 192*, 237-239 (1976).

122. Toms, G.L., Davies, J.A., Woodward, C.G., Sweet, C., Smith, H., *Br. J. Exp. Pathol. 58*, 444-458 (1977).

123. Kluger, M.J. and Vaughn, L.K., *J. Physiol. (Lond.) 282*, 243-251 (1978).

124. Vaughn, L.K. and Kluger, M.J., *Fed. Proc. 36*, 511 (1977).

125. Kluger, M.J., *in* "Environmental Physiology III, Volume 20, International Review of Physiology". (D. Robertshaw, ed.), pp. 209-251 University Park Press: Baltimore, (1979).

126. Barbour, H.G., *Physiol. Rev. 1*, 295-326 (1921).

127. Krupp, P.J. and Ziel, R., *in* "Body Temperature Regulation, Drug Effects, and Therapeutic Implications" (P. Lomax, and E. Schönbaum, eds.), pp. 383–401. Marcel Dekker, Inc., New York, (1979).

128. Kasting, N.W., Veale, W.L. and Cooper, K.E., *in* "Handbook of Experimental Pharmacology, Vol. 60" (A.S. Milton, ed.), pp. 5–24. Springer-Verlag: Berlin, Heidelberg, (1982).

129. Clark, W.G. and Coldwell, B.A., *Proc. Soc. Exp. Biol. Med. 141*, 669–672 (1972).

130. Myers, R.D., *in* "Handbook of Experimental Pharmacology: Vol. 60, Pyretics and Antipyretics". (A.S. Milton, ed.), pp. 151–186. Springer-Verlag: Berlin, Heidelberg, New York, (1982).

131. Kasting, N.W., Veale, W.L. and Cooper, K.E., *in* "Current Studies of Hypothalamic Function" (W.L. Veale and K. Lederis, eds.), pp. 63–71. S. Karger, Basel, (1978).

132. Kasting, N.W., Veale, W.L. and Cooper, K.E., *Nature 271*, 245–246 (1978).

133. Cooper, K.E., Kasting, N.W., Lederis, K. and Veale. W.L., *J. Physiol. 295*, 33–45 (1979).

134. Alexander, D.P., Bashore, R.A., Britton, H.G. and Forsling, M.A., *Biol. Neonate 25*, 242–248 (1974).

135. Zeisberger, E., Merker, G. and Blühser, S., *Pflügers Arch. 384(Suppl.)*, R26 (1980).

136. Zeisberger, E., Merker, G. and Blühser, S., *Brain Res. 212*, 379–392 (1980).

137. Merker, G., Blühser, S. and Zeisberger, E., *Cell Tiss. Res. 212*, 47–61 (1980).

138. Merker, G., Blühser, S. and Zeisberger, E., *Pflügers Arch. 384 (Suppl.)*, R25 (1980).

139. Kasting, N.W., Veale, W.L. and Cooper, K.E., *Can. J. Physiol. Pharmacol. 57*, 1453–1456 (1979).

140. Lederis, K., Pittman, Q.J., Kasting, N.W., Veale, W. and Cooper, K.E., *Biomed. Res. 3*, 1–5 (1982).

141. Eagan, P.C., Veale, W.L. and Cooper, K.E., *Proc. Can. Fed. Biol. Soc. 23*, 160 (1980).

142. Kasting, N.W., Veale, W.L. and Cooper, K.E., *Can. J. Physiol. Pharmacol. 58*, 316–319, (1980).

143. Kasting, N.W., Veale, W.L. and Cooper, K.E., *Experientia 37*, 1001–1002 (1981).

8
Clinical Presentation of 172 Heat Stroke Cases seen at Mina and Arafat—September, 1982

M. I. Al-Khawashki

Riyadh Central Hospital
Riyadh, Saudi Arabia

M. K. Y. Mustafa[1]
M. Khogali[2]

Department of Physiology,[1]
Department of Community Medicine[2]
Kuwait University, Kuwait

H. El-Sayed

King Faisal Hospital,
Taif, Saudi Arabia

The clinical presentation of 172 heat stroke cases managed over one week (25th September-2nd October 1982) at Mina and Arafat Heat Stroke Treatment Centres during the 1982 pilgrimage is reported.

The peak days of admission were 9th and 10th Dhu Al-Hijjah 1402 AH at Arafat and Mina (26th September-27th September) respectively. The peak times of admission were between 1300 and 1900 h.

The main predisposing factors were high environmental temperature, physical exertion, old age, obesity and concomitant diseases. The presenting signs were a mixture of the reported signs of both classical and exertional heat stroke. Patients presented in coma, hypotension, oliguria, diarrhoea, broncho and aspiration pneumonias and bleeding diathesis.

99

Early recovery occurred in the majority of cases. Generalized bleeding, hypotension or anuria refractory to fluid therapy and aspiration pneumonia indicated poor prognosis. Few sudden unexplained deaths occurred. In essence, every patient presented differently, which made the management more difficult although the results of use of the Makkah Body Cooling Unit (MBCU) were most rewarding.

Heat stroke is a major health problem when the annual Makkah Pilgrimage occurs during the hot cycle. Many cases of heat stroke are expected to occur. The characteristics of the pilgrim population and the operating environmental factors have been reported by Khogali (1983). The pilgrimage rituals involve strenuous physical activity from a population not acclimated to physical exercise. The majority of pilgrims come from developing countries with widespread endemic diseases. The over-crowding, noise and sanitary conditions aggravate the problem and contribute to the occurrence of heat stroke cases (Khogali, 1983).

MATERIALS AND METHODS

This report describes the clinical presentation of 172 heat stroke cases admitted and treated at two Heat Stroke Treament Centres (HSTC) at Mina and Arafat. The cases occurred during one week period, 25th September to 1st October 1982 (8th-14th Dhu Al-Hijjah 1402 AH). The peak days of admission were the 25th and 26th September at Arafat and Mina respectively. Peak times of admission were usually between 1300 and 1900 h.

Criteria for Diagnosis

The diagnosis of heat stroke was based on the following triad (Austin & Berry, 1956): (i) central nervous system disturbance; (ii) rectal temperature 40°C and above; and (iii) hot dry skin. The rectal temperature was recorded by an electronic thermometer (Digitron). All doctors involved in the diagnosis of heat stroke underwent a special training programme prior to the pilgrimage. A screening heat illness form was completed at the reception before referral to the HSTC. At the HSTC diagnosis was confirmed according to the criteria mentioned above and a complete clinical examination was performed and the presenting signs were recorded on a special form.

RESULTS

Age and Sex Distribution

Females were equal to males in number. Fig. 1 shows the age distribution of 125 cases whose ages were ascertained. Only 5 cases (3%) were under 40 years of age. The majority of cases were aged 60 years and above.

Rectal Temperature

Seventy five per cent of the patients had a rectal temperature >42°C when admitted to the HSTC. Only 12 patients (3.5%) had a rectal temperature <41°C (Fig. 2). Cooling was usually stopped when the rectal temperature dropped to 39°C.

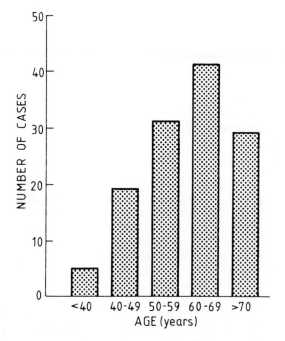

Fig. 1. Age distribution of 125 heat stroke cases seen at Mina and Arafat HSTC in 1982.

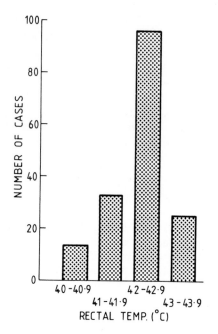

Fig. 2. Distribution of rectal temperature of 172 cases of heat stroke seen at Mina and Arafat HSTC in 1982.

During peak hours patients were removed from the MBCU with rectal temperatures of 39.5°C.

Presenting Signs

   1. *Neurological.* Typically, patients were brought in deep coma, unresponsive to painful stimuli (85%). The remaining 15% were confused, delirious and sometimes very agitated with aggressive behaviour. On examination, the pupils were found to be constricted and pin-pointed in 69% of the patients admitted. Fundus examination showed no papilloedema and none had neck rigidity. In the few patients who had lumbar puncture the CSF was clear with normal pressure. Muscle rigidity and generalised convulsions were very common especially at the start of cooling. None of the patients were observed to have signs of laterialization on admission.

2. Cardiovascular. A few patients arrived in a state of shock, gasping for breath, cyanosed with imperceptible pulse and non-recordable blood pressure. All other patients had tachycardia. Fig. 3 shows the blood pressure distribution. Twenty five percent of the patients had signs of peripheral circulatory failure as evidenced by hypotension and small volume rapid pulse. In the majority of patients the hypotension was corrected by cooling and fluid therapy. When it remained refractory to fluid therapy it rarely responded to other measures. Ectopic beats were common as detected from pulse examination since E.C.G. was not recorded on admission. Signs of congestive cardiac failure were not noted on admission but few patients had pulmonary oedema.

3. Respiratory. Tachypnoea was the rule. Difficulty in breathing was common and was due to airway obstruction caused by retention of secretions, inhalation of vomitus and in a few cases, obstruction by dentures. Fig. 4 shows signs of collapse and consolidation of the left lung secondary to

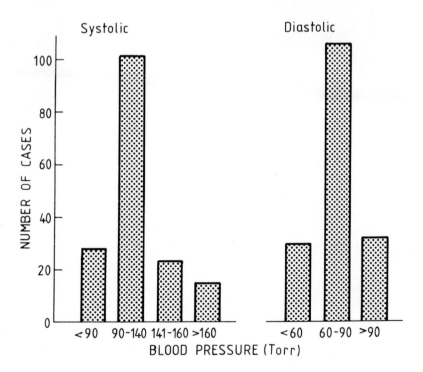

Fig. 3. Distribution of blood pressure measurements in 172 heat stroke cases seen at Mina and Arafat HSTC in 1982.

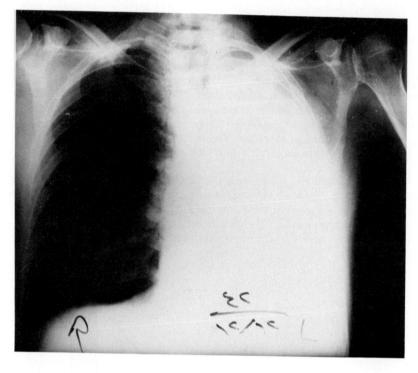

Fig. 4.   Collapse and consolidation of the left lung in a patient with heat stroke.

obstruction of a main bronchus.   Signs of bronchopneumonia and aspiration pneumonia were common.   Pulmonary oedema was seen in few cases.   Abnormalities in blood gases were very common (Mustafa *et al.*, 1983*a*).   The patients were nursed in the semi-lateral position on the MBCU, and received 28% oxygen.

4. Renal System.   Thirty three per cent of the 172 patients had anuria or oliguria defined as a urine volume of less than 500 ml after 2 litres of fluid replacement.   The majority responded to fluid replacement.   The others remained refractory to fluid therapy and in many cases to manitol infusion and frusemide.

5. *Bleeding tendencies*. Petechiael haemorrhages and ecchymosis were very common among our patients. Bleeding tendencies in the form of epistaxis, haematemesis and haematuria occurred frequently. Sometimes attempts at suction, intubation or catheterization precipitated bleeding. One female patient with a large fibroid had vaginal bleeding.

Gastrointestinal Signs

Diarrhoea was very common. It was always aggravated by cooling and continued in the post-cooling period. Stool cultures were performed on many occasions and no micro-organisms were isolated. In the MBCU the diarrhoea was easily drained while in the recovery room it was a major nursing problem.

Prognosis

Recovery was early in the majority of patients (70%). Few patients who responded favourably to cooling and fluid therapy died suddenly in the early post-cooling period. The cause of death was difficult to ascertain since post-mortems are not allowed during the Makkah Pilgrimage. Twenty per cent of patients continued in a prolonged coma and were prone to several complications. The prognosis was poor among those who had a prolonged cooling time, prolonged coma, aspiration pneumonia, generalized bleeding and those who had anuria and hypotension which remained refractory to fluid therapy. The overall mortality was 9.5% in Mina and 5% in Arafat.

DISCUSSION

Many conditions can present with hyperthermia or loss of consciousness and therefore ought to be considered in the differential diagnosis of heat stroke (Khogali, 1982). In practical terms the only problem was to differentiate heat stroke from other heat-induced illnesses. The rigid criteria applied for diagnosis of heat stroke might have resulted in under-diagnosis.

The clinical signs seen among heat stroke cases during the pilgrimage are in agreement with those reported by Austin & Berry (1956), Kew *et al.* (1967), Knochel *et al.* (1961), Knochel & Vertel (1967) and Shibolet *et al.* (1976). The

observation of constricted, pin point pupils hitherto unre-
ported could not be explained by subarachnoid haemorrhage.
It is tempting to speculate increased levels of endorphins as
the cause. Endorphins are known to be elevated under conditions of stress (Akil *et al.*, 1976). The significance of
this is whether high levels of endorphins contribute to the
coma, in which case antedotes of endorphins could play a role
in the management of heat stroke.

Differences between clinical signs of classical heat
stroke versus exertional heat stroke have been frequently
reported (Hart *et al.*, 1982; Knochel *et al.*, 1961; O'Donnel,
1975). Heat stroke victims during the pilgrimage showed
signs common in classical heat stroke in that, (a) the age
groups affected were older, (b) cases occurred in an epidemic
during heat waves, (c) patients had predisposing illnesses,
and (d) sweating was absent. On the other hand they shared
signs common to exertional heat stroke as evidenced by,
(a) the high incidence of renal failure, (b) evidence of DIC
(Mustafa *et al.*, 1983*b*) and (c) rhabdomyolysis (Gumaa *et al.*,
1983). This is to be expected since the pilgrims are an
elderly population, engaged in physical exertion while
unacclimated to physical exercise under high environmental
temperatures.

The common occurrence of respiratory complications is to
be expected since the patients were convulsive and unconscious
and definitely took a long time to reach heat stroke treatment centres. Similarly, the other factors associated with
poor prognosis (prolonged coma, hypotension and anuria refractory to fluid therapy) possibly occurred among patients
who succumbed to hyperthermia and were not attended to for a
long time. This stresses the need for prompt attention if
the mortality is to be further reduced and raises a difficult
planning question.

Acknowledgement

We are very grateful to the Ministry of Health, Saudi
Arabia and the medical staff in the Western Region for their
assistance and dedication. Our thanks to Miss Sana Jaljouli
for her help in analysis of the data.

REFERENCES

Akil, H., Madden, J., Patrick, R.L. & Barchas, J.D. (1976).
Stress induced increase in endogenous opiate peptides –
Concurrent analgesis and its partial reversal by naloxone.
In "Opiates & endogenous opioid peptides" (H.W.
Kosterlitz, ed.), pp. 63-70. Elsevier/North Holland
Biomedical Press, Amsterdam.
Austin, M.G. & Berry, J.W. (1956). Observations on one
hundred cases of heat stroke. *J. Am. Med. Assoc.*
161, 1525-1529.
Gumaa, K., Mahrouky, S.F., Mahmoud, N., Mustafa, M.K.Y. &
Khogali, M. (1983). The metabolic status of heat stroke
patients: The Makkah experience. *In* "Heat Stroke and
Temperature Regulation" (M. Khogali & J.R.S. Hales, eds),
pp. 157-169. Academic Press, Sydney.
Hart, G.R., Anderson, R.J., Crumpler, C.P., Shulkin, A.,
Reed, G. & Knochel, J.P. (1982). Epidemic classical heat
stroke, clinical characteristics and course of 28
patients. *Medicine 61*, 189-197.
Kew, J.P., Abrahams, C. & Unin, N.W. *et al.* (1967). The
effects of heat stroke on the function and structure of
kidney. *Q. J. Med. 36*, 277-300.
Khogali, M. (1982). Heat disorder with special reference to
Makkah Pilgrimage, Ministry of Health, Saudi Arabia.
Khogali, M. (1983). Epidemiology of heat illness during the
Makkah Pilgrimages in Saudi Arabia. *Int. J. Epidem.*
(in press).
Knochel, J.P., Biesel, W.R., Herndon, E.G., Gerard, L.S. &
Barry, K.G. (1961). The renal, cardiovascular, hemato-
logic and serum electrolyte abnormalities of heat stroke.
Am. J. Med. 30, 299-309.
Knochel, J.P. & Vertel, R.M. (1967). Salt loading a possible
factor in the production of potassium depletion,
rhabdomyolysis and heat injury. *Lancet i*, 659-661.
Mustafa, M.K.Y., Khogali, M. & Gumaa, K. (1983*a*). Respira-
tory pathophysiology in heat stroke. *In* "Heat Stroke and
Temperature Regulation" (M. Khogali & J.R.S. Hales, eds),
pp. 119-127. Academic Press, Sydney.
Mustafa, M.K.Y., Khogali, M. & Gumaa, K. (1983*b*). Dissemi-
nated intravascular coagulation in heat stroke. *In* "Heat
Stroke and Temperature Regulation" (M. Khogali & J.R.S.
Hales, eds), pp. 109-117. Academic Press, Sydney.
O'Donnel, T.F. (1975). Acute heat stroke – epidemiologic
biochemical, renal and coagulation studies. *J. Am. Med.
Assoc. 234*, 824-835.

Shibolet, S., Lancaster, M.C. & Danon, Y. (1976).   Heat
    stroke - A review.   *Aviat., Space & Environ. Med.*
    47, 280-301.

9
Disseminated Intravascular Coagulation among Heat Stroke Cases

M. K. Y. Mustafa[1]
M. Khogali[2]
K. Gumaa[3]

Departments of Physiology[1], Community Medicine[2]
and Biochemistry[3]
Faculty of Medicine, Kuwait University, Kuwait

N. M. Abu Al Nasr

King Faisal Hospital , Ministry of Health, Taif,
Saudi Arabia

Bleeding diathesis and shock syndrome are quite common in heat stroke. To verify if disseminated intravascular coagulation (DIC) contributes to these clinical manifestations of heat stroke, blood clotting and fibrinolysis studies were carried out on 65 non-selected heat stroke cases seen during the Makkah Pilgrimage. Ten cases (15%) were found to fulfill criteria for the diagnosis of DIC. Fourteen cases (20%) had increased fibrinolytic activity with incomplete evidence of DIC and 2 cases had severe thrombocytopenia. Cases with DIC presented with more complications and had higher mortality.

It was concluded that DIC contributes to both the bleeding manifestations and shock syndrome. The therapeutic implications are discussed.

Haemorrhagic manifestations in the form of petechial haemorrhages, ecchymosis, epistaxis and haematemesis are quite common in the clinical course of severe heat stroke (Knochel *et al.*, 1961).

The haemorrhagic state has long been attributed to increased capillary permeability, hypoprothrombinaemia and
thrombocytopenia (Wright *et al.*, 1946). Shibolet *et al.*
(1962) suggested primary fibrinolysis as the cause of the
haemorrhagic state. Disseminated intravascular clotting
(DIC) has been clearly demonstrated in cases of heat stroke
by histopathological studies which showed haemorrhages, microthrombi and focal areas of coagulative necrosis in many
organs (Malamud *et al.*, 1946; Chao *et al.*, 1981) and in a
few cases by blood coagulation studies (Meikle & Graybill,
1967; Weber & Blakely, 1969). Hart *et al.* (1982) noted
modest prolongations of prothrombin time and mild thrombocytopenia among 28 cases of classical heat stroke and suggested that DIC is only a feature of exertional heat stroke.

Also common in cases of heat stroke is the shock syndrome
(Knochel *et al.*, 1961). This is assumed to be due to peripheral circulatory failure, leading to failure of tissue
perfusion. Shock syndrome is commonly associated with DIC
which worsens the condition by the dissemination of clots in
the microcirculation.

The aim of the present study was to investigate the
occurrence of DIC in heat stroke cases seen during the pilgrimage, with a view to its importance as a contributory
factor to two features of heat stroke, viz., bleeding
diathesis and shock syndrome.

PATIENTS AND METHODS

Blood was collected in EDTA tubes for leucocyte and
platelet counts, and haematocrit estimations. Leucocyte
counts, Hb and the haemotocrit were measured by an electronic
counter. Platelets were counted in a counting chamber using
a microscope with phase contrast facility. Blood for fibrinogen and prothrombin time was collected in 3.6% citrate at
9:1 dilution. Prothrombin time was determined by the Quick
test method: 0.1 ml of patient's citrated plasma was added
to 0.1 ml rabbit brain thromboplastin (ORTHO) incubated at
37°C. The time for clot formation was measured automatically
and compared to the clotting time of human control plasma
(ORTHO).

For fibrinogen estimation, 0.2 ml of citrated plasma was
added to a tube containing 4.8 ml Owren Veronal buffer
followed by 0.2 ml of thrombin solution (fibrinodex–ORTHO).
The fibrin clot was collected by an applicator stick, rinsed
with distilled water, blotted and dissolved in NaOH.

Folin-Ciocalten reagent was added and light transmission measured spectrophotometrically and read against standardized fibrinogen concentration curves. Fibrin or fibrinogen split products (FSP) were measured semiquantitatively by the latex agglutination test (Welco Test). Two ml of blood were collected in FSP tubes. Plasma was separated and diluted 1:5 and 1:20. Latex particles were added to the serially diluted plasma. Rabbit's FSP antibodies were added. The principle of the test is that elevated levels of FSP will react with FSP antibodies and produce agglutination of the latex particles. The level of FSP was read as negative (0-10 µg/ml) when no agglutination occurred with undiluted plasma, or positive (10-40 µg/ml) when agglutination was observed in plasma with 1:5 dilution or greater than 40 µg/ml when agglutination occurred with a plasma dilution of 1:20. Positive FSP was confirmed in several samples by the ethanol gelation test (Godal & Abilgaard, 1966).

The study was carried out on non-selected heat stroke cases with or without bleeding tendencies seen at two heat stroke centres. Twenty seven patients were seen at Mina hospital in 1981 and 38 at Al-Zahir hospital in 1982 during the pilgrimage. On admission, the presenting clinical signs were recorded and blood samples were collected before commencing intravenous therapy.

RESULTS

Ten cases (15%) were found to have definite evidence of disseminated intravascular clotting as evidenced by increased FSP levels of 10 µg/ml or more together with two of the following: (i) Fibrinogen of less than 150 mg/dl; (ii) Platelet count of less than 130,000/mm^3; (iii) Prolonged prothrombin time (normal 12-14 sec).

Table I shows that 7 patients had positive FSP between 10-40 µg/ml and 3 patients had a strongly positive FSP of >40 µg/ml. Reductions in platelet counts were modest, ranging from 30 to 190 thousand with 6 patients having platelet counts of 130,000/mm^3 or less. Prolongation of the prothrombin time was very mild in the majority of patients, only being significantly prolonged in 2 patients. Fibrinogen levels ranged between 50 and 405 mg/dl, with 6 patients having fibrinogen below 150 mg/dl.

Fourteen additional patients (20%) showed increased levels of FSP, 7 at 10-40 µg/ml and 7 with FSP >40 µg/ml. Although these patients did not satisfy the full criteria for

TABLE I. *Blood coagulation studies in 10 of 65 heat stroke patients with definite evidence of disseminated intravascular clotting seen during the Makkah Pilgrimage in the years 1981 and 1982*

ID	FSP (μg/ml)	Platelet (1000/mm^3)	Fibrinogen (mg/dl)	Prothrombin Time (sec)
1982				
217	10-40	190	110	14/12
125	10-40	–	–	180/12
218	10-40	–	405	180/12
29	10-40	30	50	16/12
308	10-40	105	230	14/12
1981				
276	10-40	132	140	14/14
395	10-40	180	150	14/14
435	>40	110	70	14/14
244	>40	120	105	13/14
276	>40	132	165	14/14

FSP = fibrin-fibrinogen split products.
ID = patient identification.

DIC (Table II), one of them had a reduced fibrinogen level (140 mg/dl) and 5 had reduced platelet counts (150,000/mm^3) and probably had mild DIC. The others, besides the increased FSP levels, had normal fibrinogen levels, normal prothrombin time and normal platelet counts and could be considered to have increased fibrinolysis without consumption coagulopathy. Two patients (Table III) showed marked thrombocytopenia (25,000 and 70,000/mm^3) with normal values for FSP, fibrinogen and prothrombin time.

The clinical findings of the patients who had definite evidence of DIC were compared to those who did not (Table IV). It can be seen that the group with definite evidence of DIC were worse clinically, had lower arterial blood pH and PO$_2$, a higher incidence of oliguria or anuria and hypotension, required a longer cooling time, and had a higher mortality rate.

TABLE II. *Blood coagulation studies in 14 of 65 heat stroke patients with incomplete evidence of disseminated intravascular coagulation (DIC) seen in the years 1981 and 1982 during the Makkah Pilgrimage*

ID	FSP (μg/ml)	Platelets (1000/mm³)	Fibrinogen (mg/dl)	Prothrombin time	Remarks
40	>40	180	140	n	? DIC
205	>40	–	410	n	Fibrinolysis
243	>40	200	285	n	Fibrinolysis
53	>40	185	225	n	Fibrinolysis
166	>40	250	340	n	Fibrinolysis
153	>40	175	305	n	Fibrinolysis
176	>40	180	305	n	Fibrinolysis
50	10–40	150	–	n	? DIC
221	10–40	150	420	n	? DIC
266	10–40	148	425	n	? DIC
281	10–40	150	520	n	? DIC
353	10–40	150	375	n	? DIC
101	10–40	200	305	n	Fibrinolysis
219	10–40	180	225	n	Fibrinolysis

n = *normal prothrombin time 12-14 sec.*

TABLE III. *Blood coagulation studies in heat stroke patients with thrombocytopenia and no evidence of disseminated intravascular coagulation (DIC) seen in the years 1981 and 1982 during the Makkah Pilgrimage*

ID	FSP (μg/ml)	Platelets (1000/mm³)	Fibrinogen (mg/dl)	Prothrombin time (sec)
121	0-10	25	460	14/14
55	0-10	70	235	14/14

DISCUSSION

Disseminated intravascular coagulation with attendant fibrinolysis, is universally accepted to occur in a variety of clinical conditions. The exact mechanisms whereby DIC occurs are not known, but a number of factors are incriminated:

TABLE IV.   *Comparison of clinical findings in heat stroke patients with and without definite disseminated intravascular coagulation (DIC)*

Parameter	Definite DIC	Others
Age (yr)	60	60
	(50-70)	(35-80)
Rectal Temperature (°C)	42.6	41.9
	(42-43)	(40.8-43.2)
Coma (%)	100	60
Hypotension (%)	60	44
Anuria or Oliguria (%)		
	70	50
Arterial pH	7.19	7.27
Arterial PO$_2$ (Torr)	88	111
	(50-143)	(78-167)
Cooling Time (min)	125	78
	(60-375)	(35-185)
Mortality (%)	60	12

tissue injury and release of tissue extracts, damage to vascular endothelium, haemolysis and shistocyte formation, release of endotoxins and liberation of catecholamines in any shock syndrome (Owen *et al.*, 1975). In its most severe form, acute DIC is characterized by thrombocytopenia, hypofibrinogenaemia, depletion of factors V and VIII, and elevated levels of FSP. Fifteen per cent (10 cases) of the patients in this series satisfy most of these criteria (Table I) and support other reports of DIC in heat stroke patients (Meikle & Graybill, 1967; Weber & Blakely, 1969; O'Donnel, 1975; Chao *et al.*, 1981).

The patients with DIC showed greater association with hypoxia, acidosis and oliguria. They had prolonged cooling time and a higher mortality rate (Table IV). As with many aspects of DIC, it is difficult to separate cause from effect. Thermal injury might lead to hypercoagulability of blood, intravascular coagulation, consumption coagulopathy, bleeding diathesis and shock might ensue. On the other hand, shock from any cause is often associated with DIC and is explained by the increased levels of catecholamines which are known to shorten the clotting time (Hardaway, 1966). The common occurrence of DIC in this series of patients who had heat stroke with or without bleeding should direct attention to the additional role that DIC may play in the genesis of shock lung, renal failure and metabolic acidosis secondary to failure of tissue perfusion. Such clinical findings are common among patients with heat stroke, and widespread microthrombi and coagulative necrosis involving multiple organs was clearly shown in heat stroke deaths (Chao *et al.*, 1981).

In 14 cases (20%), complete evidence of DIC was lacking although increased levels of FSP were demonstrated (Table II). It is pertinent to note that Owen & Bowie (1978) clearly demonstrated that in dogs infused with tissue thromboplastin, the consumption of clotting factors depended on the dose and the time after infusion. Thus dogs infused with low strength thromboplastin showed a slow rate of fall of platelet counts over a few days and normal or elevated levels of prothrombin and fibrinogen. Our patients were investigated on admission, and a follow up over a few days might have revealed clearer evidence of consumption coagulopathy. On the other hand some of the cases with high levels of FSP without evidence of consumption coagulopathy might have been cases of primary fibrinolysis. Differentiation between DIC and primary fibrinolysis has an applied significance. Heparin has been successfully used in the treatment of one case of heat stroke suspected to have DIC (Weber & Blakely, 1969). Epsilon-aminocaproic acid was used for another case suspected to have primary fibrinolysis (Meikle & Graybill, 1967), which could have been harmful if the fibrinolysis was secondary to DIC.

Platelet counts in this series were generally mildly reduced. Only 2 patients showed severe thrombocytopenia (Table III). Mild thrombocytopenia could explain the common purpuric manifestations in heat stroke if accompanied by defects in platelet function. Indeed, Wright *et al.* (1946) showed that platelets submitted *in vitro* to temperatures above 42°C will not aggregate in the presence of ADP or of prothrombin. In view of the multiple defects in blood clotting mechanisms in heat stroke and the risks involved in heparin therapy we are in agreement with Shibolet & Farfel (1975) that further studies and controlled trials are needed before heparin therapy can be recommended.

In summary, we showed definite evidence of DIC in cases of heat stroke with or without bleeding tendencies seen during the pilgrimage. The DIC was associated with a complicated clinical picture and high mortality. Further studies on blood clotting, primary fibrinolysis and platelet functional defects are required.

Acknowledgement

We are very grateful to the Ministry of Health, Saudi Arabia, and the medical staff in the Western Region for the assistance and help. This work was also supported by Research Council Grant No. MC 009, Kuwait University and Kuwait Foundation for Advancement of Science.

REFERENCES

Chao, T.C., Sinniah, R. & Pakiam, J.E. (1981). Acute heat stroke deaths. *Pathology 13*, 145-156.

Godal, H.C. & Abilgaard, U. (1966). Gelation of soluble fibrin monomer complexes in human plasma. *Scand. J. Haematol. 3*, 343-350.

Hardaway, R.M. (1966). Syndromes of disseminated intravascular coagulation with special reference to shock and hemorrhage. Charles C. Thomas, Springfield, Illinois.

Hart, G.R., Anderson, R.J., Crumpler, C.P., Shulkin, A., Reed, G. & Knochel, J.P. (1982). Epidemic classical heat stroke: Clinical characteristics and course of 28 patients. *Medicine 61*, 189-197.

Knochel, C.J.P., Beisel, M.W.R., Herdon, E-G., Gerard, E.S. & Barry, K.G. (1961). The renal, cardiovascular, hematologic and serum electrolyte abnormalities of heat stroke. *Am. J. Med. 30*, 229-309.

Malamud, N., Haymaker, W. & Custer, R.P. (1946). Heat stroke a clinopathologic study of 125 fatal cases. *Milit. Surg. 99*, 397-449.

Meikle, A.W. & Graybill, J.R. (1967). Fibrinolysis and hemorrhage in a fatal case of heat stroke. *N. Eng. J. Med. 276*, 911-913.

O'Donnel, T.F. (1975). Acute heat stroke epidemiologic, biochemical, renal and coagulative studies. *J. Am. Med. Assocn. 234*, 824-828.

Owen, C.A. & Bowie, E.J.W. (1978). Induced chronic intravascular coagulation in dogs. *In* "Recent Progress in Blood Coagulation and Thrombosis Research". *Bibliotheca haemat. 44*, 169-173. Karger, Basel.

Owen, C.A., Bowie, E.J.W. & Thompson, J.H. (1975). Thrombosis and intravascular coagulation. *In* "The Diagnosis of Bleeding Disorders", pp. 284-337. Little Brown & Co., Boston.

Shibolet, S. & Farfel, Z. (1975). Heparin therapy for heat stroke. *Ann. Intern. Med. 82*, 857–858.

Shibolet, S., Fisher, S., Col, R., Gilat, T., Bank, H. & Heller, H. (1962). Fibrinolysis and hemorrhages in fatal heat stroke. *N. Eng. J. Med. 266*, 169.

Weber, M.B. & Blakely, J.A. (1969). The haemorrhagic diathesis of heat stroke: A consumptive coagulopathy successfully treated with heparin. *Lancet i*, 1190–1192.

Wright, D.O., Reppert, L.B. & Cutting, J.T. (1946). Purpuric manifestations of heat stroke: studies in prothrombin and platelets in twelve cases. *Arch. Intern. Med. 77*, 27–36.

10
Respiratory Pathophysiology in Heat Stroke

M. K. Y. Mustafa[1]
M. Khogali[2]
K. Gumaa[3]

Departments of Physiology[1], Community Medicine[2]
and Biochemistry[3],
Faculty of Medicine,
Kuwait University, Kuwait

Respiratory complications in the form of acute airway obstruction, broncho and aspiration pneumonias are quite common among cases of heat stroke seen during the Makkah Pilgrimage. Blood gas analysis was performed among 233 cases on admission to three heat stroke centres during the 1982 pilgrimage and revealed patterns of respiratory pathophysiology consistent with: primary hyperventilation (9%), hyperventilation secondary to acidosis (24%), hypoxia (40%) and hypoventilation (12%). Only 15% had normal blood gases. A high morality rate was found among the group with combined hypoxia and acidosis. Blood gas analysis proved to be essential for the management of these cases.

Respiratory complications in the form of acute airway obstruction, broncho and aspiration pneumonias are quite common among patients with heat stroke during the Makkah Pilgrimage (Khogali *et al.*, 1982). Pulmonary infarctions and wide-spread haemorrhages have been reported at postmortems (Malamud *et al.*, 1946; Levine, 1969; Chao *et al.*, 1981). Isolated clinical observations indicate the occurrence of pulmonary oedema and bronchopneumonia in heat stroke (Schrier, 1970; Shibolet, 1967). Pulmonary hyperventilation as a feature of heat stroke, has been well documented (Leithead, 1964) and supported by laboratory evidence (Austin & Berry, 1956; O'Donnel, 1975; Knochel & Caskey, 1977). Full blood

HEAT STROKE AND
TEMPERATURE REGULATION
ISBN 0 12 406180 X

gas analysis has only recently been reported (Sprung *et al.*, 1980; Khogali, 1982; Hart *et al.*, 1982). These latter reports fully characterized the acid-base status in heat stroke and Sprung *et al.* (1980) discussed the problem of hypoxia. The aim of the present investigation is to extend previous observations on blood gas analysis in heat stroke with a view to understanding the pathophysiology of respiratory complications.

METHODS

During the Makkah Pilgrimage of 1982, blood gas analysis was performed on 233 heat stroke patients at three heat stroke centres, Arafat, Mina and Al-Zahir. Arterial blood was obtained by puncture at the start of cooling, clearance of the airway and administration of oxygen by a mask which delivered 28% oxygen. The blood was immediately analysed on a blood gas analyser (ABL3 Radiometer, Copenhagen) at 37°C and automatically corrected for barometric pressure and the patient's core temperature. The corrections were done according to the method of Bradley *et al.* (1956). In general, a temperature rise of blood of 1°C will increase $PaCO_2$ and PaO_2 by 4.4% and 6% respectively while decreasing the pH by 0.0147. The analyser also read Hb concentration. Standard bicarbonate and arterial oxygen content were automatically calculated using pH, $PaCO_2$, PaO_2 and temperature values according to the Siggard-Anderson Nomogram (1963). For all the patients, clinical examination findings were recorded.

RESULTS

Of the 233 patients 40% had arterial oxygen content of 15 ml/dl or less. The low arterial oxygen content can be explained by the low oxygen tensions and the low haemoglobin concentrations; 30% of the patients were anaemic with Hb concentrations of 10 g% and less (Fig. 1). Another contributory factor to the Hb desaturation is a rightward shift in the oxyhaemoglobin dissociation curve consequent upon the hyperthermia and acidosis.

The blood gas analysis data were characterized into hyperventilation, hypoventilation and hypoxia groups. This classification (Table I) reflects categories of functional respiratory defects that throw light on the underlying respiratory

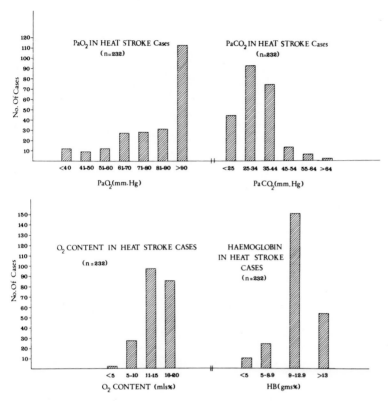

Fig. 1. Frequency distribution of PaO₂, PaCO₂, arterial oxygen content and haemoglobin concentration among 232 cases of heat stroke seen during the Makkah Pilgrimage.

pathology (Campbell *et al.*, 1974). Thirty three cases (15%) had normal blood gases and could be assumed to have no significant functional respiratory defects. Twenty cases (9%) showed a blood gas pattern characteristic of primary hyperventilation. In 58 cases (24%) the hyperventilation was secondary to metabolic acidosis. Twenty eight cases (12%) showed hypoventilation as evidenced by hypoxia and hypercapnia which in some cases approached respiratory failure. The inadequate alveolar ventilation in the latter group could be accounted for by acute airway obstruction produced by retention of secretions and inhalation of vomitus. Chest examination in these patients revealed diminished air entry and very widespread rhonchi and crepitations. Lobar or segmental collapse secondary to bronchial obstruction was

TABLE I.   Respiratory pathophysiology of 233 heat stroke patients seen in three centres during the Makkah Pilgrimage

Pathophysiology	Arafat 49 cases		Mina 117 cases		Al-Zahir 67 cases		Total 233 cases	
	No	%	No	%	No	%	No	%
Normal Blood Gases PaO$_2$>90 Torr PaCO$_2$ 35-44 Torr	9	19	17	15	7	10	33	15
Hyperventilation PaCO$_2$<35 Torr Primary (no base deficit)	4	8	12	10	4	6	20	9
Secondary to acidosis (base deficit >5 mmole/L)	10	19	36	31	12	17	58	24
Hypoventilation PaO$_2$<90 Torr PaCO$_2$>44 Torr	6	13	11	9	11	18	28	12
Hypoxia PaO$_2$<90 Torr on 28% O$_2$ PaCO$_2$<44 Torr ↑V̇O$_2$↓diffusion or V̇/Q̇ inequality	20	41	41	35	33	50	94	40

demonstrated in some cases.   In 6 cases among this group, the hypoventilation appeared to be due to depression of the respiratory centre since they had no abnormal chest findings, but were observed to have very shallow breathing.   In these 6 patients serial blood gas measurements revealed that administration of oxygen resulted in further elevation of the PaCO$_2$ and they had to be placed on assisted ventilation.   The number of patients with hypoventilation was significantly greater at Al-Zahir (18%) than at Mina (9%).

Ninety four patients (40%) were hypoxic with normal or low PaCO$_2$.   Among this group, clinical chest examination, supported by chest x-ray findings, showed frequent occurrence of broncho and aspiration pneumonias.   Both conditions are known to cause diffusion defects and ventilation/perfusion

TABLE II. *Mortality rate among 67 heat stroke patients seen at Al-Zahir Hospital related to respiratory pathophysiology*

Pathophysiology	No. of cases	No. of deaths	Morality %
Normal PaO_2>90 Torr $PaCO_2$ 33-44 Torr	7	0	0
Hyperventilation $PaCO_2$<35 Torr -Primary (no base deficit)	5	1	20

-Secondary to
  acidosis (base
  deficit
  >5 mmole/L)
	11	4	36

Hypoventilation PaO_2<90 Torr $PaCO_2$>44 Torr	11	4	36

Hypoxia
  Diffusion or
  V/Q inequality
  PaO_2<90 Torr
  $PaCO_2$<44 Torr
-No base deficit
	13	3	23

-With metabolic
  acidosis (base
  deficit
  >5 mmole/L)
	20	10	50

($\dot{V}/\dot{Q}$) inequalities. Hypoxia was also more common at the Al-Zahir centre.

The mortality rate among the 67 heat stroke cases seen at Al-Zahir centre, who had blood gas analysis, was examined in relation to the respiratory pathophysiology (Table II). This centre was selected since it had the highest overall mortality rate during the 1982 pilgrimage. It can be seen that the mortality was lower for the groups with normal blood gases (0%), primary hyperventilation (20%) and hypoxia without acidosis (23%). The highest mortality rate was seen in the group with combined hypoxia and metabolic acidosis (50%).

DISCUSSION

Clinical studies of respiratory complications supported by blood gas analysis are few. This study revealed a common occurrence of primary hyperventilation, hyperventilation secondary to acidosis, hypoxia and hypoventilation. Similar patterns, except for milder degrees of hypoxia, are apparent in the blood gas analyses of heat stroke patients reported by Sprung *et al.* (1980) and Hart *et al.* (1982). Several earlier reports have shown hyperventilation to be common in heat stroke (Leithead, 1964; Costrini *et al.*, 1979; O'Donnel, 1975). Primary alveolar hyperventilation in heat stroke can be attributed to the direct stimulating effect of hyper-thermia on the respiratory centres. Patients with this comp-lication were usually excited, restless and had a high breathing frequency, and two of them developed overt tetany. It would seem that primary hyperventilation is an early stage and is later complicated by the hypoxia and acidosis. The question of whether tissue hypoxia could be a pertinent factor in heat stroke was raised by Wyndham (1973). Results from animal experiments have shown that arterial oxygen tension remains unchanged or increased during hyperthermia (Nemoto & Frankel, 1970; Maskrey *et al.*, 1981) and human subjects exposed to hyperthermia during mild to maximal exercise showed no increase in oxygen consumption (Rowell *et al.*, 1969). The common occurrence of hypoxia in the present patients can be explained by the high rate of respiratory complications. Logistic problems made quick and safe trans-port of comatose patients to heat stroke centres unattain-able (Khogali & Al-Khawashki, 1981). Heat stroke *per se* has been shown to be associated with a high incidence of pulmon-ary infarction (Malamud, 1964; Levine, 1969), widespread pulmonary haemorrhages involving lung parenchyma (Chao *et al.*, 1981), and few cases of fatal pulmonary oedema (Vertel & Knochel, 1967; Schrier *et al.*, 1970).

Hypoxia with metabolic acidosis was found to be associ-ated with the highest mortality (Table II). This supports the finding of Hart *et al.* (1982) that cases of classical heat stroke with lactic acidosis have an unfavourable prognosis. Placement of patients in the semilateral position, which ensured a free airway and prevented the inhalation of vomitus, together with energetic suction and controlled oxygen therapy, probably contributed to the lower mortality rates observed in Mina (9.5%) and Arafat (4.7%) heat stroke centres. Blood gas analysis on admission to the heat stroke centres provides a quick, accurate appraisal of respiratory pathophysiology which is invaluable for the management of patients. Problematic cases can be quickly identified for further evaluation and close monitoring. This is essential for a situation where more than a hundred cases can be admitted over a few hours. The management can be guided by serial blood gas analysis. The hypoventilation group may need suction, intubation or assisted ventilation. The hypoxic group will require oxygen therapy and, in refractory hypoxaemia due to aspiration pneumonia, controlled positive end expiratory pressure breathing may be needed. Patients with low arterial oxygen content due to anaemia will require packed red blood cell infusion.

In conclusion, this study shows that respiratory complications reflected in abnormal blood gases, are quite common among heat stroke cases seen during the Makkah Pilgrimage. Blood gas analysis should play a major role in the management of these cases.

Acknowledgement

We are very grateful to the Ministry of Health, Saudi Arabia, and the medical staff in the Western Region for their assistance and dedication. This work was also supported by Research Council Grant No. MC 009 Kuwait University and Kuwait Foundation for Advancement of Science.

REFERENCES

Austin, M.G. & Berry, J.W. (1956). Observations on one hundred cases of heat stroke. *J. Am. Med. Assoc.* *161*, 1525-1529.

Bradley, A.F., Stupfel, M. & Severinghaus, J.W. (1956). Effect of temperature on PCO_2 and PO_2 of blood *in vitro*. *J. Appl. Physiol. 9*, 201-204.

Campbell, E.J.M., Dickinson, E.J. & Salter, J.D.H. (1974). "Clinical Physiology". English Language Book and Blackwell Scientific Publications, London.

Chao, T.C., Sinniah, R. & Pakiam, J.E. (1981). Acute heat stroke deaths. *Pathology 13*, 145-156.

Costrini, A.M, Pitt, H.A., Gustafson, A.B. & Uddin, D.E. (1979). Cardiovascular and metabolic manifestations of heat stroke and severe heat exhaustion. *Am. J. Med.* *66*, 296-302.

Hart, G.R., Anderson, R.J., Crumpler, C.P., Shulkin, A., Reed, G. & Knochel, J.P. (1982). Epidemic classical heat stroke clinical characteristics and course of 28 patients. *Medicine 61*, 169–197.

Khogali, M. (1982). Heat disorders with special reference to Makkah Pilgrimage (Hajj). Ministry of Health, Kingdom of Saudi Arabia.

Khogali, M. & Al Khawashki, M.I. (1981). Heat stroke during the Makkah Pilgrimage (Hajj). *Saudi Med. J. 2*, 85–93.

Khogali, M., Mustafa, M.K.Y. & Gumaa, K.A. (1982). Management of heat stroke. *Lancet ii*, 1225.

Knochel, J.P. & Caskey, J.H. (1977). The mechanism of hypophosphatemia in acute heat stroke. *J. Am. Med. Assocn 238*, 425–426.

Leithead, C.S. (1964). Heat illness and some related problems. *WHO Chron. 18*, 288–303.

Levine, J.A. (1969). Heat stroke in the aged. *Am. J. Med. 47*, 251–258.

Malamud, N., Haymaker, W. & Custer, R.P. (1946). Heat stroke, a clinico-pathologic study of 125 fatal cases. *Milit. Surg. 99*, 397–449.

Maskrey, M., Hales, J.R.S. & Fawcett, A.A. (1981). Effect of a constant arterial CO_2 tension on respiratory pattern in heat-stressed sheep. *J. Appl. Physiol. 50*, 515–519.

Nemoto, E.M. & Frankel, H.M. (1970). Cerebral oxygenation and metabolism during progressive hyperthermia. *Am. J. Physiol. 219*, 1784–1788.

O'Donnel, T.F. (1975). Acute heat stroke: Epidemiologic, biochemical, renal and coagulation studies. *J. Am. Med. Assocn 234*, 824–835.

Rowell, L.B., Brengelmann, G.L., Murray, J.A., Kraning II, K.K. & Kusumi, F. (1969). Human metabolic responses to hyperthermia during mild to maximal exercise. *J. Appl. Physiol. 26*, 395–402.

Schrier, R.W., Hano, J., Keller, H.I., Finkel, R.M., Gilliland, P.F., Cirksena, W.J. & Teschan, P.E. (1970). Renal, metabolic, and circulatory responses to heat and exercise. *Ann. Intern. Med. 73*, 213–223.

Shibolet, S., Col, R., Gilat, T. & Sohar, E. (1967). Heat stroke: Its clinical picture and mechanism in 36 cases. *Q. J. Med. 36*, 525–548.

Siggard-Anderson, O. (1963). Blood acid-base alignment nomogram. *Scand. J. Clin. Lab. Invest. 15*, 211–217.

Sprung, L.C., Porto-Carrero, J.C., Fernaine, V.A. & Weinberg, F.P. (1980). The metabolic and respiratory alterations of heat stroke. *Arch. Intern. Med. 140*, 665–669.

Vertel, R.M. & Knochel, J.P. (1967). Acute renal failure due to heat injury: An analysis of ten cases associated with a high incidence of myoglobinuria. *Am. J. Med.* *43*, 435–451.

Wyndham, C.H. (1973). The physiology of exercise under heat stress. *Ann. Rev. Physiol. 35*, 193–220.

11
Pulmonary Aspiration and Adult Respiratory Distress Syndrome in 40 Cases of Heat Stroke

Salah M. Soliman[1]
Zeinab AbuTaleb[1]
M. Khogali[2]
H. El-Sayed[1]

King Faisal Hospital, Taif, Saudi Arabia[1]

Faculty of Medicine, Kuwait University, Kuwait[2]

Forty heat stroke cases admitted in deep coma to the Heat Stroke Treatment Centre at Mina Hospital during two pilgrimage seasons 1401/2 AH (1981/82) are the subject of this study. Twenty cases admitted in 1981 were managed in the supine position, and the other 20 cases in 1982 were managed in the left lateral head down position (LLHD).

Fifty per cent of the 40 patients showed signs of pulmonary aspiration which varied in severity from bronchospasm to acute respiratory failure. This occurred in 65% of the patients managed in the supine position, but in only 35% of those managed in the LLHD position.

The occurrence of the clinical triad of refractory hypoxemia, decreased lung compliance and the ability to increase ventilation to maintain adequate CO_2 tension despite the increased work of breathing, make us regard this condition as an adult respiratory distress syndrome.

Ten patients with severe respiratory distress and refractory hypoxemia were mechanically ventilated with intermittent positive pressure ventilation. Six required positive end-expiratory pressures of 10-15 cm of H_2O to maintain a satisfactory O_2 tension.

On the other hand 6 patients developed acute cardiovascular failure, while 4 of them developed acute myocardial infarction in the immediate post-cooling period.

Copyright © 1983 by Academic Press Australia.
All rights of reproduction in any form reserved.

The overall mortality among the cases was 10% - and the cause of death was cardiorespiratory failure.

In conclusion the study shows that the avoidance of pulmonary aspiration and the adoption of suitable precautions during the long journey to the HSTC is the key to safety in the management of unconscious patients. These precautions should be strictly adhered to at the site of collapse, during transport, in the reception and during cooling.

Heat stroke is a clinical condition characterized by a triad of anhidrosis, disturbance of consciousness and a temperature above 40°C. It occurs in epidemic proportions during the Makkah Pilgrimage (when in the hot weather) and if not effectively treated, will carry a high mortality rate of up to 80%.

Heat stroke can produce a reversible impairment of brain functions which may be lethal if not managed early. It can also produce foci of permanent damage leading to permanent sequelae such as paraparesis, cerebellar ataxia, dysarthria, polyneuropathy or dementia.

Several clinical manifestations which represent multiple-system damage have been noticed in patients with heat stroke, including tremors, convulsions, excessive salivation, vomiting, watery diarrhoea, tachypnoea, pulmonary oedema, pulmonary infarction, tachycardia, hypotension, bleeding tendency, metabolic acidosis, hypokalaemia, hyperglycaemia and acute renal failure.

The onset of heat stroke can be sudden and the loss of consciousness may be preceded by a period of a general weakness, headache, dizziness, restlessness, delerium or mental confusion, before collapse of the patient occurs.

In our experience, respiratory distress in comatosed patients suffering from heat stroke is a major problem. This paper describes a study conducted at the Heat Stroke Treatment Centre (HSTC) at Mina Hospital during the 1401 AH and 1402 AH (1981 and 1982) Hajj to throw some light on this complex problem.

PATIENTS AND METHODS

From the comatose patients admitted to the HSTC at Mina Hospital during the Hajj, 20 were selected from Hajj 1401 (Group 1) and 20 from Hajj 1402 (Group 2), i.e., a total of 40 comatose patients are included. There were 28 males and

12 females (one 7 mth pregnant), of 40 to 80 yrs age, and many nationalities; 20 were obese but no weight was recorded.

On clinical examination all patients were comatose with a rectal temperature (T_{re}) of 40-43.4°C. The skin was hot and dry and 7 patients showed signs of dehydration.

On admission, blood and urine samples were taken before cooling, after cooling and after 12 and 24 h. The venous blood was sent to the laboratory for the recommended investigations of electrolytes, urea, creatinine, glucose, SGOT, SGPT, LDH, haemoglobin, haemotocrit, platelets, leukocytes, bleeding time, coagulation time, prothrombin time and activity (Khogali *et al.*, 1983). Arterial blood samples were sent for blood gas analysis (ABL 3, Radiometer, Copenhagen). Chest x-rays were performed daily to assess the radiological changes in the lungs.

Once the diagnosis of heat stroke was confirmed clinically, cooling started on the Body Cooling Unit (BCU) (Khogali, 1983). When body temperature had dropped to 38.5°C cooling was discontinued and the patient transferred to the post-cooling recovery section where the vital signs, pulse, temperature, blood pressure, respiratory rate and urinary output were recorded. The C.V.P. in the critically ill patients and in patients with dehydration was also measured (Khogali & Weiner, 1980).

Group 1 patients were managed in the supine position in the pre-cooling, cooling and post-cooling period, with an airway and oxygen mask. Group 2 was managed in the left lateral with head down position during the same stages with an airway and oxygen mask. Frequent and effective suction of secretions and vomitus were performed in both groups.

In 5 comatose patients with respiratory complications, endotracheal intubation was performed during cooling, which enhanced suction of the tracheobronchial tree.

RESULTS

The mean and standard deviation of the values for blood gases and acid-base analysis of the 40 comatosed patients performed on admission is shown in Table I.

Half the 40 patients included in this study exhibited signs of pulmonary aspiration which varied in its severity from bronchospasm to acute respiratory insufficiency to pulmonary oedema. Out of these 20 patients, 13 were among those managed in the supine position during Hajj 1401 while

TABLE I.   *Mean values of Blood Gas and Acid-base Status of 40 patients admitted during Hajj 1401 and 1402*

Measurement	Group 1		Group 2	
	Mean	± SD	Mean	± SD
T_{re} (°C)	41.6	0.9	42.1	1.0
Hb (g)	12.5	1.7	12.7	2.3
pH	7.33	0.1	7.28	0.12
$PaCO_2$ (Torr)	37.0	9.9	34.2	6.4
PaO_2 (Torr)	114.7	48.5	120.1	49.2
HCO_3^- (m eq/l)	17.8	3.2	15.1	4.5
BE (m eq/l)	-5.2	4.1	-8.9	6.5

TABLE II.   *The Incidence of Pulmonary Aspiration among the 40 comatose patients*

	No. of patients	Position	Pulmonary aspiration	%
Group 1	20	Supine	13	65%
Group 2	20	Left lateral head down	7	35%
Total	40		20	50%

the remaining 7 patients were from those managed in the left lateral and head down position (Table II).

Clinically, dyspnoea, cyanosis, tachypnoea and/or respiratory distress were observed. Chest auscultation elicited scattered wheezes, rhonchi and crepitations.

Gas exchange in the lungs deteriorated over 2 to 3 days as evidenced by blood gas analysis. Chest x-ray showed patchy opacities which were unilateral in 5 cases and bilateral in 15 cases. When the lung condition deteriorated further, additional opacification occurred and the lung field became opaque. Blood gas analysis showed that hypoxaemia increased with deterioration of the clinical and radiological findings of the lungs.

Of the 20 patients with pulmonary aspiration, 10 developed severe respiratory distress and refractory hypoxaemia as evidenced by PaO_2 of less than 50 Torr; of these, 6 had normal $PaCO_2$, 2 had low $PaCO_2$ and 2 high $PaCO_2$.

The 10 patients were mechanically ventilated with intermittent positive pressure ventilation (IPPV) with high oxygen concentrations (60-100%) in order to increase their PaO_2 to a satisfactory level. From those ventilated with IPPV, 6 patients required a positive end-expiratory pressure (PEEP) of 10 to 15 cm H_2O in order to improve their PaO_2.

On the other hand, 6 patients from those with acute respiratory insufficiency developed acute cardiovascular failure. Two of them developed multiple ventricular extrasystoles and 4 developed acute myocardial infarction in the immediate post-cooling period; 2 of the 4 required controlled respiration with IPPV and dobutamine drip to support both the respiratory and cardiovascular systems.

A variety of pulmonary complications were observed in the other 10 patients ranging from mild bronchospasm to bronchopneumonia which improved on medication; chest x-rays and blood gas analyses showed marked improvement in subsequent days.

DISCUSSION

Pulmonary complications occur more frequently in comatose patients, especially in the obese and elderly with pre-existing lung disease. These patients are more at risk, and awareness of the pending complication and adoption of proper precautions are a key to safety.

The unconscious patient is also at risk from any partial airway obstruction. This might be due to the tongue falling back or due to pulmonary aspiration of vomitus, blood and secretions. The entry of solid matter to the bronchial tree may obstruct one of the larger bronchi.

While any aspirated fluid may obstruct the airway and interfere with normal ventilation/perfusion patterns, the chemical burn induced by acid aspiration appears to be the most harmful agent. Mendelson (1946) found that acid was the main cause of pulmonary injury seen after aspiration, and that it is rapidly distributed throughout the lungs. This causes a loss of alveolar capillary integrity and an exudation of fluid and protein into the interstitial spaces, alveoli and bronchi. It is the pH of the aspirate that determines the severity of damage to the lungs. A pH of 3 or less associated with aspiration pneumonitis and other unknown factors influence the outcome of the condition. As a rule, the larger the volume of aspirate the poorer the prognosis (Awe *et al.*, 1966).

The exudation causes pulmonary oedema, a decrease in the pulmonary compliance, an increase in lung weight and a significant intrapulmonary perfusion shunting with subsequent hypoxaemia. After aspiration, the intrapulmonary shunt may be as much as 50% of the cardiac output (Cameron *et al.*, 1973). Sufficient fluid may be lost from the circulation to deplete intravascular volume and aggrevate hypotension.

Pulmonary aspiration of gastric contents may cause any of the following: atelectasis, bronchospasm, pneumonitis, pulmonary oedema, pulmonary necrosis and haemorrhage, hypoxaemia, acidosis and circulatory collapse. The reported mortality ranges from 35 to 77% (Reines *et al.*, 1981). However, not all who aspirate suffer damage to the lungs, while sometimes there may be a mild bronchopneumonia with only slight systemic upset.

The most immediate and severe physiologic problem is the hypoxaemia that occurs within minutes of the aspiration, secondary to reflex airway closure in response to the aspiration of acidic fluid.

Destruction or alteration of normal surfactant activity and the outpouring of fluid and protein into the damaged tissues causes interstitial oedema which results in further pulmonary embarrasment. Thus the aspiration of a significant amount of stomach contents into the lungs causes in its worst form, an acute respiratory failure regardless of the nature of the aspirate.

The $PaCO_2$ is either normal or low after acid aspiration. On the other hand the $PaCO_2$ may be higher than normal after foodstuff aspiration, indicating a great degree of obstruction and hypoventilation.

Hypoxaemia produces pulmonary hypertension through pulmonary vasoconstriction. Hypoxic pulmonary vasconstriction may contribute to right heart failure in patients with respiratory insufficiency. These patients need a therapeutic agent which decreases pulmonary vascular resistance and increases cardiac contractility. Oxygen reduces the pulmonary vascular resistance but requires a relatively long period to become fully effective.

Furman *et al.* (1982) studied hypoxic pulmonary vasoconstriction in the pig and concluded that dobutamine and not dopamine is inhibitory. Dopamine is a valuable agent in the treatment of cardiac failure, but does not offer the potential advantage of pulmonary vasodilatation and may behave as a

pulmonary vasoconstrictor. On the other hand, if dobutamine
proves to be effective as a pulmonary vasodilator during
hypoxia in man (as it has in pigs), its relative freedom from
adverse side effects may make it an attractive agent for the
treatment of hypoxic pulmonary vasoconstriction and right
side heart failure.

Hypotension, hypoxaemia, hypokalaemia, acidosis and hypoxic pulmonary vasoconstriction, all contribute to a variety
of disturbances in myocardial functions and may lead to acute
cardiovascular failure, especially in those patients who are
often elderly and already have myocardial pathology.

On the other hand, hypoxia, hypercarbia, acidosis and a
high venous pressure are all harmful to the recently damaged
brain. This risk starts at the site where the patient
develops his illness and unless steps are taken to prevent it,
the patient will remain at risk during the time spent in the
ambulance, emergency reception or even during cooling in the
Body Cooling Unit and in the post-cooling ward.

Following pulmonary aspiration, there frequently occurs a
clinical triad characterized by: refractory hypoxaemia,
decreased lung compliance and the ability to increase ventilation transiently to maintain adequate PCO_2 despite the increased work of breathing; we therefore regard this condition
as Adult Respiratory Distress Syndrome (ARDS).

In the absence of clinically identifiable causes of true
shunting (e.g., acute lobar or segmental atelectasis, consolidation pneumonitis, vascular lung tumours and cardiac right
to left shunt), it is reasonable to assume that a refractory
hypoxaemia may be due to an ARDS.

From this study, it was noticed that for patients with
coma persisting even after cooling, mechanical suction of
secretions, mucus, vomitus, and foreign bodies and positioning in the LLHD position, with a simple airway and oxygen
mask, are needed to decrease the risk of pulmonary aspiration.

Any clinical sign of airway obstruction or arterial blood
gas abnormalities should be followed by immediate endotracheal
intubation with a cuffed tube, and the patient should be
managed by an assisted or controlled ventilation. If the
patient cannot tolerate the tube or begins to fight the ventilation, sedation or even a muscle relaxant is indicated to
facilitate controlled ventilation. Improvement in blood gas
can be achieved by the use of IPPV.

In our study, 10 patients (25% of the cases) were suffering from severe respiratory distress and refractory hypoxaemia with PaO_2 less than 50 Torr. They were mechanically
ventilated with IPPV. Of these, 6 patients required a PEEP
of 10 to 15 cm H_2O to increase the PO_2 to a satisfactory level

(around 100 Torr).   In adult patients, 3 to 8 cm H_2O PEEP
should not be considered therapeutic for acute restrictive
pathology.

On the other hand, an increase in body temperature
together with acidosis and electrolyte disturbance may lead
to terminal cardiac failure and this may occur in heat stroke
as well as in malignant hyperpyrexia (Ellis, 1972).   Six
patients from our series developed acute cardiovascular failure, together with acute respiratory insufficiency.   Of these,
2 showed frequent ventricular extrasystoles and were managed
by correction of body temperature and the associated hypokalaemia and metabolic acidosis, together with infusion of
Xylocain.   The other 4 patients developed acute myocardial
infarction of various severity in the immediate post-cooling
period.   Two of them required controlled ventilation and
dobutamine infusion to support both the respiratory and
cardiovascular systems.

A variety of pulmonary complications were also observed
in another 10 patients of our series, ranging from bronchospasm to pulmonary oedema, and were managed by routine medical
procedures which gave good results as evidenced by clinical,
radiological and blood gas analysis.

CONCLUSION

Although rapid and effective cooling is the most important step in the management of patients with heat stroke, it
must go hand-in-hand with other measures needed to correct
any respiratory, cardiovascular, fluid and electrolyte and
acid-base abnormalities.

After the occurrence of heat stroke, the treatment object-
ives must also be directed towards the avoidance of secondary
injury to the recently damaged brain from hypotension, hypoxaemia, hypercarbia and acidosis.

In many cases of pulmonary aspiration the stomach contents
can be avoided, which is preferred whenever possible.   The
frequency of pulmonary aspiration in the unconscious patient
should be reduced by careful observation and by positioning
in the LLHD position.   This management should start at the
site of collapse, and continue during transport and in the
emergency reception at hospital.   The care rendered by the
general doctors and the emergency medical personnel at the
site of the accident and during transport, has a highly
significant impact on the ultimate outcome.

The clinical and radiological changes of the lungs in the advanced form of pulmonary aspiration are indistinguishable from that in ARDS.

Intrapulmonary shunting may be so severe that even the administration of 100% oxygen does not achieve an acceptable PO_2. Mechanical ventilation with IPPV and associated manouvres, are usually instituted to improve gas exchange, and a PEEP of 10 to 15 cm H_2O should be considered when more than 50% oxygen is required to maintain a satisfactory PO_2 during IPPV. On the other hand the medical management should include chest physiotherapy, tracheobronchial toilet, treatment of infection and appropriate fluid therapy. Also, measures to decrease lung water, re-expand collapsed lungs and decrease airway resistance are needed.

Acknowledgement

We are most grateful to the Minister of Health, Saudi Arabia, and to the medical and nursing staff without whom this work would never be completed.

REFERENCES

Awe, W.C., Fletcher, W.S. & Jacob, S.E. (1966). The pathophysiology of aspiration pneumonitis. *Surgery, St. Louis* *60*, 232–238.

Cameron, J.L., Mitchell, W.H. & Zuidema, G.D. (1973). Aspiration pneumonia: Clinical outcome following documented aspiration. *Archs. Surg., Chicago 106*, 49–52.

Ellis, F.P. (1972). Mortality from heat illness and heat-aggrevated illness in the United States. *Environ. Res.* *5*, 1–58.

Furman, W.R., Summer, W.R., Kennedy, T.P. & Sylvester, J.T. (1982). Comparison of the effects of dobutamine, dopamine and isoproterenol on hypoxic pulmonary vasoconstriction in the pig. *Crit. Care Med. 10*, 371–374.

Khogali, M. (1983). The Makkah Body Cooling Unit. *In* "Heat Stroke and Temperature Regulation" (M. Khogali & J.R.S. Hales, eds), pp. 139–148. Academic Press, Sydney.

Khogali, M. & Weiner, J.S. (1980). Heat stroke: Report on 18 cases. *Lancet ii*, 276–278.

138 *Salah M. Soliman et al.*

Khogali, M., ElSayed, H., Amar, M., Sayad, S.E., Al Habashi, S. & Mutwalli, A. (1983). Management and therapy regimen during cooling and in the recovery room at different heat stroke treatment centres. *In* "Heat Stroke and Temperature Regulation" (M. Khogali & J.R.S. Hales, eds), pp. 149-156. Academic Press, Sydney.

Mendelson, C.L. (1946). Aspiration of stomach contents into lungs during obstetric anesthesia. *Am. J. Obstet. Gynec. 52*, 191-205.

Reines, H.D., Spicer, K.M., Cooper, J.F. & Redding, J.S. (1981). Use of Technetium-99 labelled albumin to detect capillary leak. *Crit. Care Med. 9*, 148.

12
The Makkah Body Cooling Unit

M. Khogali

Department of Community Medicine,
Kuwait University, Kuwait

*Thermal cellular injury and circulatory changes associat-
ed with heat stroke result in widespread tissue injury in the
heart, kidney, liver and in the blood coagulation system.
The severity of damage is determined by both the degree and
duration of hyperthermia; rapid treatment is critical.*

*Traditionally, cooling is by immersion in a tub of iced
water with the skin massaged vigorously to minimize cutaneous
vasoconstriction. This form of cooling has physiological,
practical and hygienic disadvantages. Vasoconstriction
impedes transfer of heat from the body core to surface. It
is practically difficult to administer oxygen, fluids and
monitor blood pressure with the patient in a tub of water. A
tub of water mixed with vomitus and diarrhoea of comatosed
patients is unhygienic to both the patient and attendants.*

*The Makkah Body Cooling Unit - especially designed to
overcome the above-mentioned shortcomings - is described here.
Cooling is achieved through evaporation of atomised spray
from a warm skin. By keeping the skin warm it is possible to
maintain a high water-vapour pressure gradient from skin to
air, thereby facilitating heat loss, eg., a drop in rectal
temperature of 0.3°C per 5 min. The mobile net system,
ensures maximum exposure of skin and easy transport of heat
stroke patients, who are commonly very obese. Semilateral
positioning of the patient to guarantee a free airway is also
facilitated. Moreover monitoring, administering of oxygen
and fluids and other aspects of management are easily performed without interruption to cooling.*

*This work was supported by Research Council Grant No.
MC 009, Kuwait University.*

HEAT STROKE AND
TEMPERATURE REGULATION
ISBN 0 12 406180 X

139

The seriousness of uncontrolled heat-induced pyrexia lies particularly in the damage it can do to the central nervous system. Heat stroke patients suffer muscular spasms and tremor, delirium, convulsions, vomiting and defaecation. The muscular activity also adds to body heat production and thus aggrevates the already high body temperature. The damage to the nervous system will depress and damage the respiratory centres and the patient will become comatosed and eventually die from respiratory failure.

Thus the heat stroke patient, if he is to be saved, needs to have the hyperpyrexia reduced rapidly, and at the same time he must be given treatment for the neurological and respiratory disturbances. In many cases, particularly the elderly, pre-existing and predisposing disorders must also be corrected. In fact, it does not take long either to boil an egg or to cook neurones. The importance of prompt cooling of heat stroke patients to below 39°C cannot be over-emphasised (Hamilton, 1976).

METHODS OF COOLING

The time-honoured methods are to cool the patient by (1) immersion in a bath of ice-cold water (i.e. a mixture of ice and water), (2) by packing in ice, (3) by sponging with ice-cold water, (4) sponging with cold water combined with blowing room air over him, or (5) spraying with cold water combined with air movement. These methods have many physiological, hygienic and practical objections. These include:

(1) Intense peripheral vasoconstriction with shunting of blood away from the cooling surface (the skin) with a possible paradoxical increase in the core temperature. The vasoconstriction impedes the transfer of heat from the body core to the surface - which is an essential process in eliminating body heat.

(2) Shivering, induced by the cold skin, raises metabolic heat production, which in the hyperthermic state is already two to three times the basal level (Bradbury *et al.*, 1967).

(3) Extreme discomfort to the semiconscious and conscious patient with resulting agitation and struggling which further increases metabolic heat production and the oxygen requirements of the tissues (Lancet, 1982).

(4) Extreme discomfort for the medical attendants who have to massage the patient in the ice-water bath.

(5) Difficulty performing cardiopulmonary resuscitation and other required treatment appropriate to the clinical condition of the patient.

(6) Monitoring of the body temperature and other indices, like blood pressure, become very difficult, the patient has to be lifted repeatedly out of the tub and dried before any measurement is taken.

(7) If profuse diarrhoea and vomiting occur while the patient is in the tub, the operation becomes extremely unpleasant and unhygienic to the attendant and dangerous to the patient.

EXPERIMENTAL WORK

On the request of Ministry of Health, Saudi Arabia, Weiner and Khogali of the London School of Hygiene and Tropical Medicine began their search for a new method which could overcome the undesirable physiological and practical shortcomings outlined above. The hypothesis of a new method based on evaporative cooling from a warm skin was put forward. Scientifically and physiologically it was sound but needed experimental evidence.

A series of comparative experiments were planned. Human volunteers dressed in plastic suits were rendered hyperthermic and then cooled by one of the five following procedures in a room kept at 20–22°C dry bulb, and 17–19°C wet bulb temperatures (Weiner & Khogali, 1980):

(i) Control – The hyperthermic volunteer lay on a net suspended over a specially designed bath tub; no cooling was performed.

(ii) Water mattress – The hyperthermic volunteer lay on a water-filled mattress at 20°C.

(iii) Water bath – The hyperthermic volunteer was immersed, except for the head, in a stirred water bath kept at 15°C.

(iv) Cold air spray – The hyperthermic volunteer lay on a net suspended over a bath; atomised water at 15°C was sprayed continually from above and below while air at room temperature was blown over the subject to evaporate from the wet surface.

(v) Warm air spray – The hyperthermic volunteer lay on a net and atomised water at 15°C was continually sprayed while warm air at 45°C, reaching the subject at 33°C, was blown over the subject.

The above-mentioned experiments proved clearly that the Warm Air Spray method was by far the most efficient as well as acceptable to the patients. The difference in cooling efficiency of the methods was highly significant (Weiner & Khogali, 1980). The skin temperature was kept at the range 30-34°C which prevented shivering and enhanced physiological evaporation.

To prove the high efficiency of the method our human volunteers were cooled while exercising (Fig. 1) and their oxygen uptake being measured. The objective was to measure the rate of cooling at a high oxygen uptake level similar to heat stroke patients. The results were very satisfactory.

THE MAKKAH BODY COOLING UNIT

Following the success achieved by the comparative experimental studies, Weiner and Khogali designed the Makkah Body Cooling Unit (MBCU) (manufactured by the Western Medical Group, London, England). The BCU MK I was tested using human volunteers and gave the expected results.

The BCU MK II was then tested during the 1978 pilgrimage in Saudi Arabia when only three heat stroke cases were admitted and successfully treated. Many modifications were introduced following observations during field work which resulted in the highly modified MBCU MK VI (Fig. 2).

Between 1979 and 1982, 1597 cases of heat stroke have been successfully treated in Saudi Arabia (Table I).

The Biomechanical Basis of the MBCU

The MBCU consists of two main components, the cooling bed and the service cabinet-console. The bed is in the shape of a tub-bath with a strong stretched net which slides on the rims of the beds. It is provided with two sets of ducts, one below and the other above the patient. Heated air from the console is blown through circular holes of 1.25 cm diameter spaced equidistantly to allow the uniform flow of air which is blown at very low pressure. The atomized spray is jetted through nozzles from a tank of water with the supply

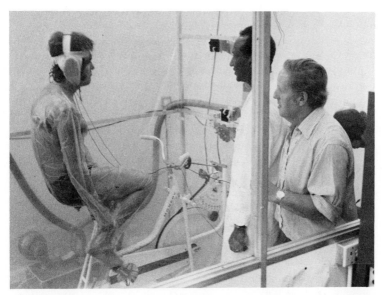

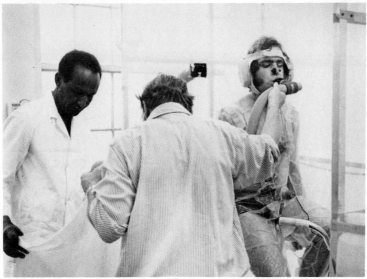

Fig. 1. Human volunteer rendered hyperthermic then cooled by warm air spray and oxygen uptake measured while exercising.

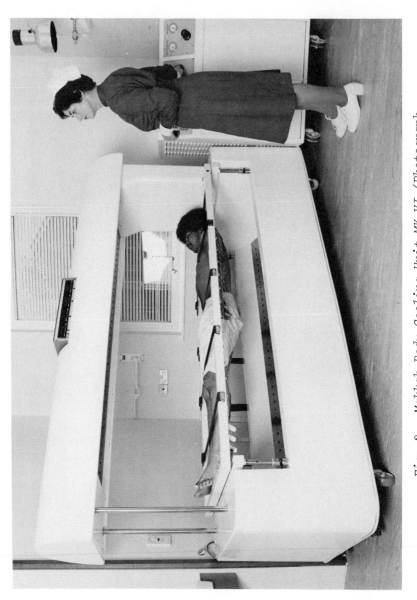

Fig. 2.   Makkah Body Cooling Unit MK VI (Photograph courtesy of the Western Medical Group, London, England).

incorporated in the compressed air line. Compressed air is supplied from a compressor directly linked to the console or from a central compressed air supply.

The bed has a drainage outlet at the foot end. Straps are available for securing the patient if restless.

The service cabinet comprises (1) a control panel to regulate the flow and temperature of air and to regulate the pressure of the spray which can be operated either automatically or manually, and (2) a fan for the supply of heated air.

The MBCU operates on straightforward, basic, physiological principles (Khogali & Weiner, 1980). Rectal and several skin surface temperatures are monitored using thermistors and a "Digitron" system. The patient's heat is dissipated by evaporation of atomized water from a warm skin. Keeping the skin warm maintains vasodilatation, which ensures a high level of blood flow and heat flow from the body core to the skin. The warm, carefully wetted skin simulates sweating and provides a high vapour pressure which facilitates the removal of water vapour. The high rate of impinging air flow also increases the total volume of water vapour and heat removed.

The cooling power of the MBCU is approximately three times that of the average person sweating at his highest rate.

TABLE I. Heat Stroke Cases treated using MBCU during 1979-82

Place	No. of Centres	Heat stroke cases treated			
		1979 Nov.	1980 Oct.	1981 Oct./Sept.	1982 Sept.
Makkah	3	?	95	293	709
Mina/Arafat	2	18	32	104	222
Medina	1	?	49	81	188
Total	6	18	176	478	1119

The Medico-Clinical Aspects of the MBCU

While rapid and effective cooling is the corner-stone of treatment of heat stroke, it is important during and after cooling to treat shock, correct fluid and electrolyte disturbances, and support vital organs. Close monitoring of the patient entails venepuncture for laboratory tests, blood pressure measurements, ECG recording and urine collections for laboratory tests. Treatment of the derangements brought about by hyperpyrexia, particularly convulsions, coma and respiratory failure needs easy access to the patient. The MBCU is designed in such a way as to enable all these operations to be carried out effectively and hygienically. Access to the patient by the medical attendants is possible from both sides and both ends.

CONCLUSION

The MBCU is designed in such a way as to enhance maximal evaporative cooling of the body. The cooling rate achieved by the MBCU is extremely efficient. Analysis of the core body temperatures of patients treated by the MBCU shows a drop of 0.3°C every 5 min (Fig. 3). This is a very high physiological cooling rate considering that heat stroke patients have a very high metabolic heat production (Al-Khawashki *et al.*, 1983). Sheep rendered hyperthermic to a critical level of heat stroke temperature and then cooled on the MBCU at the same intermittent cycle used for patients, exhibited the same cooling rate as achieved in patients (Mustafa *et al.*, 1983). The data also show that the maximum cooling rate is achieved in 30 min which is a life-saving procedure (Fig. 3).

Acknowledgement

I am grateful to the Minister of Health, Saudi Arabia, for his generous support and to the late Professor J.S. Weiner, Professor of Environmental Physiology, University of London. My thanks to all who contributed to the development of the MBCU and to Dr M. Attia for his help in analysis of the data.

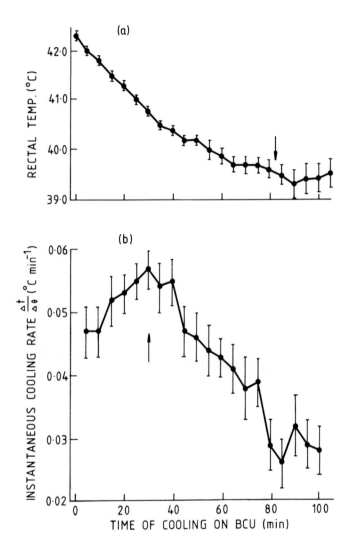

Fig. 3. Cooling of 69 heat stroke patients on the MBCU: (a) Rectal temperature decrease; cooling stopped when rectal reached 39.5°C (arrow). (b) Instantaneous cooling rate, with point in time of maximum cooling rate indicated.

REFERENCES

Al-Khawashki, M.I., Khogali, M., El Ergsus, A.M. & Titchiner, J. (1983). The biomedical principles of a body cooling unit for treatment of heat stroke. *J. Biochem. Engng.* (in press).

Bradbury, P.A., Fox, R.H., Goldsmith, R., Hampton, I.F.G. & Muir, A.L. (1967). Resting metabolism in man at elevated body temperature. *J. Physiol. 189*, 61-79.

Hamilton, D. (1976). The immediate treatment of heatstroke. *Anaesthesia 31*, 207-272.

Khogali, M. & Weiner, J.S. (1980). Heat stroke - Report on 18 cases. *Lancet ii*, 276-278.

Lancet (1982). Leading article. Management of heat stroke. *Lancet ii*, 910-911.

Mustafa, M.K.Y., Khogali, M., Elkhatib, G., Attia, M., Gumaa, K., Nasralla, A., Mahmoud, N.A. & Al Adnani, S.A. (1983). Sequential pathophysiological changes during heat stress and cooling in sheep. *In* "Thermal Physiology" (J.R.S. Hales, ed.)(in press). Raven Press, New York.

Weiner, J.S. & Khogali, M. (1980). A physiological body cooling unit for treatment of heat stroke. *Lancet i*, 507-509.

13
Management and Therapy Regimen during Cooling and in the Recovery Room at Different Heat Stroke Treatment Centres

M. Khogali[1]
H. El-Sayed[2]
Mustafa Amar[3]
Saad El Sayad[4]
Sayed Al Habashi[5]
Azzam Mutwali[2]

Faculty of Medicine, Kuwait University, Kuwait[1]

King Faisal Hospital, Taif[2]

AlZahir Hospital, Makkah[3]

Kings Hospital, Medina[4]

AlShisha Hospital, Makkah, Saudi Arabia[5]

Successful management of heat stroke cases requires early detection, prompt diagnosis, prompt physiologically effective cooling, continuous monitoring and frequent physical examination, meticulous respiratory care and fluid replacement in the comatose patient.

Heat Stroke Treatment Centres (HSTC) have been established at Arafat, Mina, AlZahir, Agiad and AlShisha hospitals in Makkah and at Kings Hospital, Medina. The treatment regimens followed in all six centres is the result of wide field experience and takes into consideration all points of successful management.

The results of management based on the mortality rates of 1982 varied significantly: Agiad (4.7%), Arafat (7.5%), Mina (9.5%), AlShisha (13.4%), Medina (14%) and AlZahir (18.4%). The total number of cases managed during a 3 week period was 1119 cases, ranging between 53 at Arafat and 321 at Alzahir.

HEAT STROKE AND
TEMPERATURE REGULATION
ISBN 0 12 406180 X

The 222 cases at Arafat and Mina hospitals were managed during a 5 day period. The other centres received patients over a 3 to 4 week period.

Major determinants of the prognosis for heat stroke patients are the level of body core temperature and time lapse before cooling. Accessibility of the HSTC, emergency care during transport and prompt diagnosis at reception in hospitals and health centres can explain the differences observed in mortality rates. Trained staff and good clinical judgement may also contribute.

The regimens followed and differences in each HSTC are discussed with a view to striving to attain zero mortality.

Heat stroke must be recognized as a medical emergency. Inappropriate delay in diagnosis or treatment may lead to death or irreversible organ damage (Sprung, 1980). Between 1979 and 1982, approximately 1800 cases were treated at the differant   Heat Stroke Treatment Centres (HSTC) along the pilgrimage route (Fig. 1). The dramatic increase in numbers from 18 cases in 1979 to 1119 in 1982 is due at least partly to the cyclic movement (into hotter months) of the annual Makkah Pilgrimage (Khogali, 1983*a*).

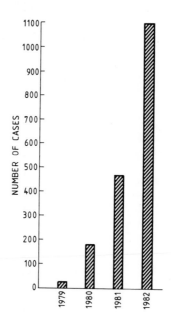

Fig. 1. Heat stroke cases treated in all HSTC during 1979-82 (1399-1402 AH).

HEAT STROKE TREATMENT CENTRES (HSTC)

HSTC have been established at different hospitals. For each Makkah Body Cooling Unit (MBCU) (Khogali, 1983*b*) there are 2 to 3 recovery beds and at least 8 post-cooling beds. Each HSTC comprises a number of MBCUs. The efficiency of the MBCU in treatment of heat stroke cases has been reported (Khogali & Weiner, 1980; Khogali & AlKhawashki, 1981). Adjacent to the cooling room, there is a recovery room which is well staffed and under supervision of the staff of the HSTC. After recovery the patients are transferred to post-cooling wards to be kept under supervision and for follow-up.

In 1982, Makkah, the holy city which houses the holy Kaaba had three HSTCs - one at Agiad Hospital, which is adjacent to the holy Kaaba, with four MBCUs; one in AlZahir Hospital (the main hospital), with four MBCUs; and the third in AlShisha Hospital, with three MBCUs.

The other two centres along the pilgrimage route are situated in Mina General Hospital, with five MBCUs and at Arafat General Hospital with four MBCUs. The pilgrims are supposed to stay for one day (9th Dhu Al Hijja AH) at Arafat, and for three days (10-12th Dhu Al Hijja AH) at Mina in camps. One centre is situated at Kings Hospital, Medina, 450 km north of Makkah. Medina houses the tomb of Prophet Mohamed, and pilgrims visit it either prior to or after the holy pilgrimage (Khogali, 1983*a*).

HEAT STROKE CASES

The number of heat stroke cases managed at each centre during the three year period 1980-1982 (1400-1402 AH) is shown in Fig. 2. In 1980 only 176 cases were admitted to the HSTC and in 1981 the number increased to 467 cases and jumped to 1119 cases in 1982. Of these 1119 cases, 709 (63%) were admitted to three centres at Makkah, 321 at AlZahir Hospital, 254 at Agiad Hospital and 134 at AlShisha Hospital. These cases were seen over a period of 4 weeks from November 25 to December 23, 1402 AH. At Arafat 53 cases were admitted on December 8 and 9, 1402 AH, and at Mina 169 cases were admitted between the 8th and the 13th day. The 177 cases seen at Medina were admitted over two periods, November 15 to

December 1, 1402 AH and from December 12 to 25. The chronological occurrence of cases is important because it is closely related to the rites of the pilgrimage and the pilgrims'
movement (Khogali, 1983*a*).

MANAGEMENT OF HEAT STROKE CASES

The procedure for detection and referring cases to an HSTC
has been outlined (AlMarzoogi *et al.*, 1983). On arrival at
the HSTC the protocols and procedure of management are stand-
ardised. This was done with a view to maintaining a unified
system of management and care. This did not interfere with
the personal clinical judgement of the doctors in charge to
prescribe any treatment they thought appropriate at the time.

PROCEDURE OF MANAGEMENT

The detailed regimen to be followed has been outlined by
Khogali (1982). The salient features of the programme are
the following:-

(i) Check rectal temperature and confirm diagnosis.
(ii) Conduct a clinical examination and enter all information on the clinical examination forms provided.
(iii) If confirmed as a case of heat stroke, put the
patient immediately on the MBCU in a semilateral position,
head down, and start cooling.
(iv) Keep the airways clear and administer oxygen.
Apply suction if needed.
(v) Sedation: Give valium 10 mg intravenously and
repeat the dose as necessary. If the patient shows severe
convulsions give largactil and phenergan 50 mg each in 250 ml
of 25% glucose, divided into two doses.
(vi) Establish an indwelling catheter in the bladder and
collect urine; record volume and send a sample to the
laboratory.
(vii) Blood samples: Withdraw 0.5 ml arterial blood for
blood gas analysis, and 10-15 ml venous blood for the laboratory investigations shown in Table I.
(viii) Intravenous fluids: The first three litres should
be administered very quickly, starting with normal saline.
Continue fluids, depending on the condition of the patient,
the response to treatment and laboratory investigations. If

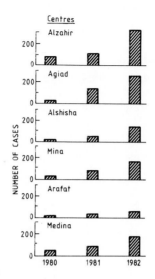

Fig. 2. Annual distribution of heat stroke cases by location of treatment 1980-82 (1400-1402 AH).

there is oliguria (after hydration of the patient) give 500 ml of 10% mannitol. If no response, consult a nephrologist.
    (ix) Stop cooling when rectal temperature drops to 38.5°-39°C and remove the patient to the recovery room.

MANAGEMENT AT THE RECOVERY ROOM

    The recovery room is adjacent to the cooling room. The management of the patient is continued depending on his condition and results of the laboratory tests. Repeated samples of arterial blood for gas analysis are taken. Venous blood is withdrawn after 6, 12 and 24 h for the same battery of investigations listed in Table I. The metabolic status of the heat stroke patients is reported by Gumaa *et al.* (1983). From the recovery room the patient is transferred to the post-cooling ward. Time taken in the recovery room ranged between 1 and 24 h.

154   *M. Khogali et al.*

TABLE I.   List of Recommended Laboratory Investigations Performed at the HSTC

A. *Biochemical:*

1. *Aspartate Transaminase (SCOT)*
2. *Alanine Transaminase (SGPT)*
3. *Lactate Dehydrogenase (LDH)*
4. *Creatine Phosphokinase (CPK)*
5. *Serum Calcium*
6. *Serum Chloride*
7. *Serum Potassium*

8. *Serum Sodium*
9. *Serum Bilirubin*
10. *Serum Creatinine*
11. *Blood Glucose*
12. *Blood Urea*
13. *Serum Lactate*

B. *Haematological:*

1. *Haematocrit*
2. *Haemoglobin*
3. *Leucocytes*
4. *Platelets*

5. *Bleeding Time*
6. *Coagulation Time*
7. *Prothrombin Time*

C. *Others:*

1. *Blood Film for Malaria*

CHARACTERISTICS OF PATIENTS AND CLINICAL PRESENTATION

There was a slight predominance of males over females among our patients. Obesity was very common and besides its medical problems, it added to the difficulty of transporting the patients. The age distribution was similar to that reported by Mustafa *et al.* (1983), the majority being aged 50 years and more. The clinical presentation in all six centres was also similar to that reported by Mustafa *et al.* (1983), with 70–80% presented in deep coma. The range of body temperature recorded was 40°–43.2°C at Mina and Arafat, 40°–46.2°C at Medina and 40°–46.5°C at Makkah.

RESULTS OF MANAGEMENT

The results of management based on the mortality rates varied significantly form one centre to another. The lowest mortality was experienced at Agiad Hospital (4.7%) and the highest mortality at AlZahir Hospital (18.4%). Arafat, Mina, AlShisha and Medina Hospitals experienced mortality rates of 7.5%, 9.5%, 13.4% and 14%, respectively.

The major complications were neurological, particularly
cerebellar dysfunction, confusion, hemiparesis, amnesias and
personality changes. Proper follow-up of these patients was
very difficult. Once slightly improved, most of them decide
to continue their pilgrimage whatever the consequences.
Those who go back to their own countries are difficult to
trace. It is very difficult to assess the exact sequel of
events following the post-heat stroke status of a number of
patients. Those who recovered fully were in good normal
condition when discharged.

DISCUSSION

Each one of our patients was burdened with one or more
of the predisposing factors of heat stroke. The majority
were in fact burdened with several predisposing factors as
well as concommittant chronic diseases.

Time taken to reach the HSTC varied greatly from one
centre to another. It could have been minutes in Agiad and
Arafat Hospitals or hours in the case of AlZahir and Mina.
The absolute level of core temperature and the time the
patient remains at that level are the two most important
factors in the outcome of heat stroke (Lind, 1983). In our
view the low mortality at Arafat and Agiad centres is very
largely attributable to their accessibility and consequently
the early management of cases. Moreover at Arafat, most
pilgrims are still active and just at the beginning of the
strenuous rites of the pilgrimage.

At Agiad hospital a new factor was added. Every patient
admitted to the HSTC was given dexamethasone with a view to
relieving any brain oedema. The mortality is significantly
lower but not due to dexamethazone administration. A controlled clinical trial needs to be conducted to verify this
claim.

In conclusion the regimen recommended and adopted in the
HSTC during the previous three years was successful. During
1982, due to the serious complication of bleeding (especially
epistaxis) we introduced at Mina HSTC, low-molecular-weight
dextran (500 ml) as part of the fluid regimen that is given
to patients during the first three hours (Khogali *et al.*,
1982). Dextran is a volume expander which improves renal
flow and may help prevent sludging of platelets.

Acknowledgement

Our gratitude and thanks to the Minister of Health, Saudi Arabia, who initiated the idea. Our thanks to the Medical personnel and staff of the health services who with their dedication and co-operation made this plan successful.

REFERENCES

AlMarzoogi, A., Khogali, M. & ElErgesus, A. (1983). Organizational set up for detecting, screening, treatment and follow up of heat disorders. *In* "Heat Stroke and Temperature Regulation" (M. Khogali & J.R.S. Hales, eds), pp. 31-39., Academic Press, Sydney.

Gumaa, K., Al-Mahrongi, S., Mahmoud, N.A., Mustafa, M.K.Y. & Khogali, M. (1983). The metabolic status of heat stroke patients: The Makkah experience. *In* "Heat Stroke and Temperature Regulation" (M. Khogali & J.R.S. Hales, eds), pp.157-169. Academic Press, Sydney.

Khogali, M. & Weiner, J.S. (1980). Heat stroke: Report on 18 cases. *Lancet ii*, 276-278.

Khogali, M. & AlKhawashki, M.I. (1981). Heat stroke during the Makkah Pilgrimage. *Saudi Med. J. 2*, 85-93.

Khogali, M. (1982). Heat disorders. A monograph - published by the Ministry of Health, Saudi Arabia.

Khogali, M. (1983*a*). Epidemiology of heat illnesses during the Makkah Pilgrimages in Saudi Arabia. *Int. J. Epid.* (in press).

Khogali, M. (1983*b*). The Makkah Body Cooling Unit. *In* "Heat Stroke and Temperature Regulation" (M. Khogali & J.R.S. Hales, eds), pp.139-148. Academic Press, Sydney.

Khogali, M., Mustafa, M.K.Y. & Gumaa, K. (1982). Management of heat stroke. *Lancet ii*, 1225.

Lind, A.R. (1983). Pathophysiology of heat exhaustion and heat stroke. *In* "Heat Stroke and Temperture Regulation" (M. Khogali & J.R.S. Hales, eds), pp.179-188. Academic Press, Sydney.

Mustafa, M.K.Y., Khogali, M. & Gumaa, K. (1983). Respiratory pathophysiology in heat stroke. *In* "Heat Stroke and Temperature Regulation" (M. Khogali & J.R.S. Hales, eds), pp.119-127. Academic Press, Sydney.

Sprung, C.L. (1980). Heat stroke: Modern approach to an ancient disease. *Chest 77*, 461-462.

14
The Metabolic Status of Heat Stroke Patients: The Makkah Experience

K. Gumaa[1]
S. F. El-Mahrouky[2]
N. Mahmoud[3]
M. K. Y. Mustafa[3]
M. Khogali[4]

Departments of Biochemistry[1], Physiology[3]
and Community Medicine[4]
Faculty of Medicine,
Kuwait University, Kuwait

Baljurashy Hospital[2], Taif, Saudi Arabia

One hundred and forty nine heat stroke patients had their metabolic and acid-base status evaluated on admission, and in 94 cases again immediately after cooling. The pattern observed was related to the accessibility of treatment centres. Thus, in Arafat compensated respiratory alkalosis predominated (30.6%) while in Mina compensated metabolic acidosis (34.5%) was the most frequent presenting derangement. Body cooling and rehydration had little effect on the acid-base status, correction of which should be attended to aggressively in the post-cooling period.

Most patients were hyponatraemic and normokalaemic and all were hyperglycaemic on admission. Rehydration aggravated the hyponatraemia, precipitated hypokalaemia but decreased the hyperglycaemia. Raised blood urea and creatinine were common presenting features that reverted to normal on rehydration. There was a tendency for plasma enzymes to be raised with no specific organ pattern, confirming that multi-system damage occurs in heat stroke.

The metabolic and respiratory changes in heat stroke patients have been reported for small groups of patients from the USA during the heat waves of the summers of 1977 and 1978 (Sprung *et al.*, 1980; Hart *et al.*, 1982). Although mixed disorders were common, the most frequent presenting acid-base disturbance was respiratory alkalosis secondary to heat-induced hyperventilation (Gaudio & Abramson, 1968). All patients were hypokalaemic at some point in their course (Hart *et al.*, 1982).

Unlike previous reports, heat stroke during the Makkah Pilgrimage involves a heterogeneous muslim population that arrives from all over the world by rapid means of transportation which do not allow for acclimatization. Most pilgrims are over 50 years old, an age-group with prevalence of chronic diseases, eg. diabetes mellitus, cardiovascular and respiratory diseases (Khogali, 1982). Many pilgrims are known to go into voluntary dehydration that may be extreme in hot summers in order to avoid the use of the overcrowded sanitary facilities. Added to these factors are the physical exertion involved in the performance of the rites, and the air and noise pollution, all acting in concert to promote a stressful state.

During the 1982 Makkah Pilgrimage, 149 of the heat stroke cases admitted for treatment in Arafat and Mina hospitals were investigated biochemically on admission, and the tests repeated in the immediate post-cooling period in 94 of the patients seen in Mina hospital.

METHODS

Arterial blood for blood gas analysis was drawn anaerobically from the radial, brachial or femoral artery and immediately analysed in an ABL-3 analyser (Radiometer, Copenhagen). Blood gas values were automatically corrected for the patient's core temperature (Severinghaus, 1966) and their acid-base status evaluated with the help of a triangular acid-base diagram (Instrumentation Laboratory (UK) Ltd). Venous blood for lactate concentration was drawn into an equal volume of 1.0 M perchloric acid, mixed and analysed by the method of Marback & Weil (1967). Serum electrolytes, glucose, urea, creatinine, lactate dehydrogenase, and alanine and aspartate aminotransferases were determined on an auto-analyser (Astra-8, Beckman) using standard methods.

All data are expressed as the mean ± S.E.M. and statistical significance analyzed by Student's t-test.

RESULTS AND DISCUSSION

A. Acid-Base Status

The admission arterial blood gases, pH, bicarbonate and base excess are shown in Table I. The most frequent presenting acid-base status was compensated metabolic acidosis (Mac. C. 31%), and in decreasing order of frequency, uncompensated metabolic acidosis (Mac. 20.1%), compensated respiratory alkalosis (Ral. C. 19.5%), normal status (N. 10.7%), respiratory alkalosis (Ral. 8%), respiratory acidosis (Rac. 4%), combined metabolic and respiratory acidosis (Mac. Rac. 4%), metabolic alkalosis (Mal. 2%) and combined metabolic and respiratory alkalosis (Mal. Ral. 0.7%).

The apparent divergence of the present findings from the reported preponderance of respiratory alkalosis in heat stroke (Hart *et al.*, 1982) may possibly be due to the variable time intervals between succumbing to heat stroke and arrival at treatment centres. Analysis of our data according to the accessibility of the hospital from the site of pilgrimage activity, revealed that the most frequent acid-base abnormality in heat stroke patients on the first day of the pilgrimage was compensated respiratory alkalosis (30.6%) with uncompensated and compensated metabolic acidosis at 25% and 22% respectively (Fig. 1). This pattern prevailed in Arafat where the hospital is accessed with relative ease. In contrast, the admission acid-base status on the second to the

TABLE I. Arterial Blood Gases, Bicarbonate and Base Excess in 149 Heat Stroke Patients on Admission

Status	n	pH	PCO_2 (Torr)	HCO_3^- (mmol/l)	BE (mmol/l)	PO_2 (Torr)
Mal. Ral.	1	7.52	32	25	4.1	73
Mal.	3	7.47±0.01	37±1	26±2	3.2±1.8	63±0.4
Mac. Rac.	6	7.18±0.03	57±5	19±2	-7.6±1.3	79± 9
Rac.	6	7.32±0.04	55±3	26±2	1.2±2.2	125±37
Ral.	12	7.46±0.01	30±1	21±0.5	-0.7±0.3	97±15
N.	16	7.39±0.01	39±1	22±0.3	-1.0±0.3	88±12
Ral. C.	29	7.45±0.01	25±1	16±1	-4.3±0.6	114±11
Mac.	30	7.30±0.01	39±1	18±0.5	-6.4±0.6	108±12
Mac. C.	46	7.35±0.01	27±1	14±0.5	-9.0±0.6	109± 6

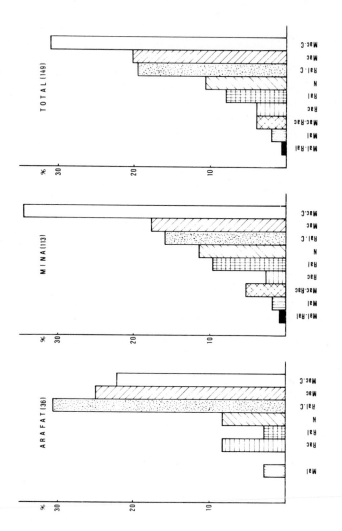

Fig. 1.   The acid-base status of heat stroke patients on admission.

fourth day of the pilgrimage was similar to the pattern observed for the whole group, since these patients represented 76% of those investigated.

Extrapolations from the above observations suggest that the initial derangement upon succumbing to heat stroke is acute respiratory alkalosis, precipitated by the heat-induced hyperventilation (Gaudio & Abramson, 1968). This transient condition is rapidly compensated by the setting in of metabolic acidosis, the progress of which reflects the severity of preceding physical exertion, dehydration, hypotension and/or tissue hypoxia, all of which promote the development of lactic acidosis (Hart *et al.*, 1982). However, the picture may be complicated by pre-existing diseases. In those cases where blood lactate concentration was determined, the maximal elevation was 6.35 mM yet the overall mean was 1.7 ± 0.09 mM (n = 100). These values suggest that in most cases seen, the heat stroke was not exertional but classical. In these cases initial lactate values greater than 3.3 mM are indicators of poor prognosis (Hart *et al.*, 1982).

The effect of body cooling and rehydration on the acid-base status was examined in 94 heat stroke cases treated in Mina hospital. No significant alteration in the overall picture was noted (Fig. 2). However, the number of cases with normal status and those with metabolic acidosis increased with a corresponding decrease in those with respiratory alkalosis and those with compensated metabolic acidosis.

B. *Sodium, Potassium and Glucose Metabolism*

Heat acclimation is classically described as a triad of effector actions: increased sweat rate, decreased body temperature and reduced heart rate (Henane, 1980). Sweat secretion may reach large volumes of more than 12 l daily depending on the atmospheric temperature and physical activity (Robinson & Robinson, 1954) and may be a major route for losses of electrolytes (Knochel *et al.*, 1972). Modern, fast means of travel would not allow for acclimation and therefore pilgrims would be expected to develop electrolyte deficits of varying degrees depending on climatic conditions prevailing in the countries of origin.

Table II reveals that Nigerians, Iraqis & Jordanians, and Turks & Syrians with heat stroke presented with mean plasma $[Na^+]$ not significantly lower than normal, while others were hyponatraemic. Since no correction for the dehydration was made, the deficit may be underestimated. Cooling and rehydration of patients lowered the $[Na^+]$ further

TABLE II. *Plasma Sodium, Potassium and Glucose in 94 Heat Stroke Patients on Admission (Adm.) and in the immediate Post-Cooling Period (P.C.)*

Nationality	n	Sodium (mEq/l)		Potassium (mEq/l)		Glucose (mmol/l)	
		Adm.	P.C.	Adm.	P.C.	Adm.	P.C.
N Africans	14	125 ± 7	-	5.8 ± 1.9	-	14.4 ± 1.4	-
Nigerians	3	133 ± 8	-	3.9 ± 0.1	-	11.2 ± 3.2	-
Egyptians	14	131 ± 2	129 ± 4	4.3 ± 0.3***	2.8 ± 0.2	17.5 ± 2.2***	8.6 ± 1.6
Iraq & Jord.	3	133 ± 8	-	4.9 ± 1.2	-	17.4 ± 8.1	-
Indonesians	7	125 ± 6	-	3.2 ± 0.3	-	11.6 ± 2.1	-
Turks & Syr.	8	139 ± 6	131 ± 4	3.8 ± 0.4*	2.9 ± 0.1	12.6 ± 3.1	-
Indians	7	128 ± 6	143 ± 4	4.2 ± 0.9	2.6 ± 0.1	12.9 ± 1.9	7.1 ± 1.1
Unknowns	38	130 ± 1	-	4.3 ± 0.6	-	11.8 ± 0.7	10.2 ± 2.3

$*P < 0.05$; $***P < 0.001$.

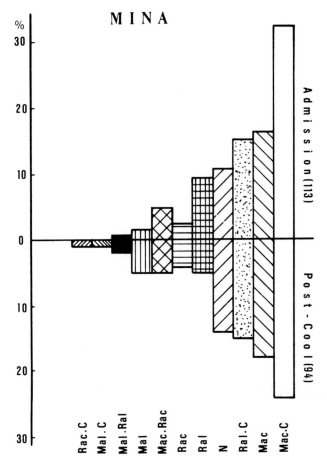

Fig. 2. The effect of body cooling and rehydration on the acid-base status of heat stroke patients.

except in Indians in whom the hyponatraemia was corrected.

Hyperkalaemia was evident in North Africans while others were normokalaemic on admission. However, cooling and rehydration resulted in hypokalaemia in all subjects investigated.

All patients were significantly hyperglycaemic and in 17 of them there was concomittant glycosuria on admission. Egyptians were the most hyperglycaemic (blood glucose 17.5 ± 2.2 mM) while Nigerians were the least hyperglycaemic (blood glucose 11.2 ± 3.2 mM). Cooling and rehydration decreased the glucose concentration, though not to normal fasting levels and the possibility that some patients were diabetic has not been excluded.

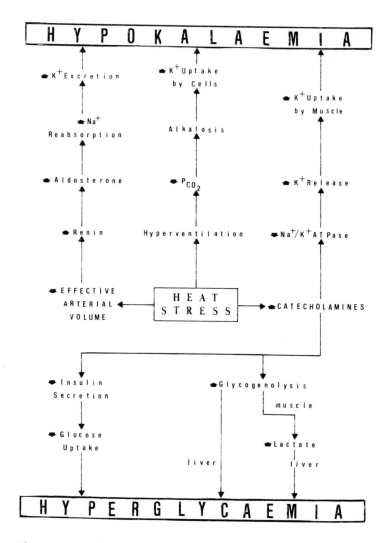

Fig. 3.   Sodium, potassium and glucose metabolism in heat stress.

To reconcile these observations, it appears that heat stress, through stimulation of sweating, results in losses of Na⁺ and water. The resulting hypovolaemia (Fig. 3) would increase renin release (Escourrou *et al.*, 1982) which, through angiotensin, increases aldosterone secretion (Knochel *et al.*, 1972) and thus increases Na⁺ reabsorption by the kidneys in exchange for K^+. Simultaneously, heat-induced hyperventilation decreases the P_{CO_2}, and the resulting alkalosis shifts K^+ into the intracellular compartment, thus potentiating the hypokalaemia. Furthermore, heat stress increases the sympathetic release of catecholamines (Escourrou *et al.*, 1982) which induce a biphasic change in plasma $[K^+]$. There is an initial hyperkalaemia due to release of K^+ from the liver and splanchnic organs through inhibition of the Na^+/K^+ ATPase (Tria *et al.*, 1974), followed by re-uptake by the liver and by muscle (Vick *et al.*, 1972). In addition, catecholamines inhibit insulin release from the pancreas and therefore decrease glucose uptake by peripheral tissues and cause hyperglycaemia. They also stimulate glycogenolysis in liver and muscle, the latter contributing to lactate production which augments the observed hyperglycaemia through hepatic gluconeogenesis unless splanchnic vasoconstriction sets in, when lactic acidosis will prevail. When the heat stress is relieved by cooling and rehydration, insulin secretion is resumed, promoting glucose and K^+ entry into peripheral tissues, with consequent lowering of their plasma concentrations.

C. *Renal Function*

The mean plasma urea concentration on admission was 3.2 ± 0.37 mM for Arafat hospital, and 7.1 ± 0.56 mM for Mina hospital. The corresponding values for creatinine were 0.15

TABLE III. *Plasma Urea and Creatinine in 94 Heat Stroke Patients on Admission (Adm.) and in the immediate Post-Cooling Period (P.C.)*

Nationality	n	Urea (mmol/l)		Creatinine (mmol/l)	
		Adm.	P.C.	Adm.	P.C.
N Africans	14	6.0± 0.6	–	0.13±0.01	–
Nigerians	3	7.3± 0.2	–	0.19±0.04	–
Egyptians	14	10.7± 3.4	6.1±1.0	0.36±0.15	0.13±0.02
Iraq & Jord.	3	4.0±12.9	–	0.16±0.55	–
Indonesians	7	7.1± 0.7	–	0.13±0.01	–
Turks & Syr.	8	6.0± 0.5*	9.2±1.5	0.13±0.02	0.13±0.04
Indians	7	5.8± 1.7	6.8±1.2	0.16±0.04	0.16±0.04
Unknowns	38	6.5± 0.4	–	0.16±0.02	–

* $P < 0.05$

± 0.01 mM and 0.21 ± 0.04 mM respectively. These findings demonstrate that, the longer the interval between the onset of heat stroke symptoms and admission, the more deranged is the renal function. Among the patients seen, Egyptians suffered most (Table III) but reverted to the normal range upon cooling and rehydration. In contrast, Turks and Syrians presented with normal urea and creatinine levels but cooling and rehydration then resulted in a higher urea concentration with no corresponding change in creatinine, probably reflecting a hypercatabolic state.

Thus, it appears that transient changes in renal function in heat stroke are secondary to hypovolaemia and are therefore reversed by rehydration.

D. *Plasma Enzymes*

Table IV demonstrates that there was a trend for lactate dehydrogenase (LDH), alanine aminotransferase (ALT) and aspartate aminotransferase (AST) to be raised with no particular pattern in heat stroke patients. A rise in the blood level of all diagnostic enzymes was reported for similar cases and attributed to extensive multisystem damage (Shibolet *et al.*, 1967). The elevation in LDH activity may result from rhabdomyolysis, haemolysis or other tissue damage. We could not unequivocally attribute the rise to rhabdomyolysis since no creatine phosphokinase activity or urine myoglobin determinations were available. Furthermore, all five isoenzymes of LDH, determined in a pilot study, were elevated with no specific organ profile.

Cooling and rehydration lowered plasma enzyme activities except in Indians in whom there was a significant rise of AST. The present findings confirm the experimental observations of Magazanick *et al.* (1981) that blood enzyme levels at the point of collapse in heat stroke are of no diagnostic value, though elevations observed 4-5 h later may give a fair indication of the extent of tissue damage.

CONCLUSION

The value of laboratory investigations in the management of heat stroke cases should not be underestimated. Evaluation of the acid-base status and electrolyte determinations should constitute the front-line tests necessary to provide guidance for fluid therapy. However, correction of the acid-base

TABLE IV. *Plasma Enzymes in 94 Heat Stroke Patients on Admission (Adm.) and in the immediate Post-Cooling Period (P.C.)*

Nationality	n	LDH (IU/l) Adm.	LDH (IU/l) P.C.	ALT (IU/l) Adm.	ALT (IU/l) P.C.	AST (IU/l) Adm.	AST (IU/l) P.C.
N Africans	14	219± 49	–	22± 4	–	43±11	–
Nigerians	3	308	–	29±24	–	44±10	–
Egyptians	14	230± 65	307±18	54±14	38±13	83±23	60±13
Iraq & Jord.	3	198±207	–	28±51	–	66±83	–
Indonesians	7	203± 57	–	60±36	–	94±50	–
Turks & Syr.	8	149± 27	102±48	33± 5	48±23	46±11	27±12
Indians	7	193± 28[a]	64±11	48±24	19± 4	36±23[a]	131± 4
Unknowns	38	142± 11	–	34± 6	–	44± 7	–

[a] $P < 0.001$

status may be deferred until after the core temperature is lowered to levels compatible with survival, and thereafter attended to aggressively outside the urgent atmosphere prevailing in the body cooling areas. Renal function tests, though important for monitoring progress, can be regarded as second-line tests since estimates of renal function may be gauged from monitoring urine output relative to fluid infused. The last line of tests is constituted by plasma enzymes which should yield useful information regarding the extent of damage persisting in the post-cooling period and therefore aid in planning the course of management during the recovery phase.

Acknowledgement

We are very grateful to the Ministry of Health, Saudi Arabia, and the medical staff in the Western Region for their assistance and dedication. This work was also supported by Research Council Grant No. MC 009 of Kuwait University and Kuwait Foundation for Advancement of Science.

168   *K. Gumaa et al.*

REFERENCES

Escourrou, P., Freund, P.R., Rowell, L.B. & Johnson, D.G. (1982). Splanchnic vasoconstriction in heat stressed men: role of renin–angiotensin system. *J. Appl. Physiol.* *52*, 1438-1443.

Gaudio, R. Jr. & Abramson, N. (1968). Heat-induced hyperventilation. *J. Appl. Physiol.* *25*, 742-746.

Hart, G.R., Anderson, R.J., Crumpler, C.P., Shulkin, A., Reed, G. & Knochel, J.P. (1982). Epidemic classical heat stroke: Clinical characteristics and course of 28 patients. *Medicine 61*, 189-197.

Henane, R. (1980). Acclimatization to heat in man: Giant or windmill. A critical reappraisal. *In* "Contributions to Thermal Physiology" (Z. Szelényi & M. Székely, eds), pp. 275-284. Akadémiai Kiadó, Budapest.

Khogali, M. (1982). Heat disorders with special reference to Makkah Pilgrimage (Hajj). Ministry of Health, Kingdom of Saudi Arabia.

Knochel, J.P., Dotin, L.N. & Hamburger, R.J. (1972). Pathophysiology of intense physical conditioning in a hot climate. I. Mechanisms of potassium depletion. *J. Clin. Invest. 51*, 242-255.

Magazanik, A., Epstein, Y., Shapiro, Y. & Sohar, E. (1980). Effects of heatstroke, dehydration and exercise on blood enzyme levels in the dog. *In* "Contributions to Thermal Physiology" (Z. Szelényi & M. Székely, eds), pp. 549-552. Akadémiai Kiadó, Budapest.

Marback, E.P. & Weil, M.H. (1967). Rapid enzymatic measurement of blood lactate and pyruvate. *Clin. Chem. 13*, 314-325.

Robinson, S. & Robinson, A.H. (1954). Chemical composition of sweat. *Physiol. Rev. 34*, 202-220.

Severinghaus, J.W. (1966). Blood gas calculator. *J. Appl. Physiol. 21*, 1108-1116.

Shibolet, S., Col, R., Gilat, T. & Sohar, E. (1967). Heatstroke: Its clinical picture and mechanism in 36 cases. *Q. J. Med.36*, 525-548.

Sprung, C.L., Portocarrero, C.J., Fernaine, A.V. & Weinberg, P.F. (1980). The metabolic and respiratory alterations of heat stroke. *Arch. Intern. Med. 140*, 665-669.

Tria, E., Luly, P., Tomasi, V., Trevisani, A. & Barnabei, O. (1974). Modulation by cyclic AMP *in vitro* of liver plasma membrane ($Na^+ - K^+$)-ATPase and protein kinases. *Biochim. Biophys. Acta 343*, 297-306.

Vick, R.L., Todd, E.P. & Luedke, D.W. (1972). Epinephrine-induced hyperkalaemia: Relation to liver and skeletal muscle. *J. Pharmacol. Exp. Ther.* *181*, 139-146.

15
Clinical Picture and Management of Heat Exhaustion

Mohamed Al-D'bbag[1]
M. Khogali[2]
Mahmoud Ghallab[3]

King Faisal Hospital, Taif[1]

Directorate of Health Affairs, Makkah, Saudi Arabia[3]

Department of Community Medicine,[2]
Faculty of Medicine, Kuwait University, Kuwait

Heat exhaustion is the commonest clinical disorder during the Makkah Pilgrimage when it takes place during the heat of summer. Thousands of cases have been managed at the health centres in Mina and in the main hospitals along the pilgrimage route, with no details recorded. Specially designated wards and a protocol of management were in operation. There was great difficulty in differentiating between salt-depletion and water-depletion heat exhaustion per se. *The high prevalence of pyrexia and thirst favours diagnosis as "water-depletion heat exhaustion", but due to frequent upper respiratory infection our patients were managed as cases of mixed heat exhaustion. Both isotonic saline and 5% dextrose were administered, combined with evaporative cooling from wet gauze.*

Of 347 cases treated over 7 days at Mina Hospital, 242 (70%) were males and 105 (30%) females. Forty six per cent were aged 40-59 and 24% were <40 yrs old. Of the 36 nationalities affected, the largest group (97 cases, 28%) were Egyptians. Body temperature mean was 39.2°C, but it exceeded 41°C in 14 cases. Ninety two per cent of those with blood pressure records had normal blood pressure. The average stay in the treatment ward was 50 min. Twenty per cent complained of weakness, 14% of giddiness and 13% of fatigue. Twenty four per cent presented with hot dry skin.

The main shortcomings, viz., lack of supportive laboratory investigations and enough trained staff, are rectifiable.

HEAT STROKE AND
TEMPERATURE REGULATION
ISBN 0 12 406180 X

171

Heat exhaustion is the most common clinical disorder to prevail during the Makkah Pilgrimage when it occurs during high environmental temperature. With 2 million population at risk, many cases of heat exhaustion are expected. In fact, thousands have been managed at different medical institutions along the pilgrimage route.

Pilgrims camp at Mina for 3-5 days. The medical facilities provided at Mina comprise Mina General Hospital, Ministry of Health, health centres strategically located among the pilgrims' camps and the services provided by the Islamic Medical Missions.

Here we report our experience in management of cases of heat exhaustion seen at Mina General Hospital.

PATIENTS AND METHODS

During the 1402 AH (1982) pilgrimage two wards, each with 16 canvas beds, were designated for cases of heat exhaustion at the reception at Mina General Hospital.

Three hundred and forty seven cases were admitted to these wards during a one week period, 6th-12th Dhu Al-Hijja 1402 AH (24-30th September 1982). Of these patients 223 (64%) were admitted on the 10th and 11th of Dhu Al-Hijja 1402 AH, 104 (30%) were admitted on the 8th and 12th while the remaining 6% were treated on the 6th and 7th. No cases were admitted on the 9th, since all pilgrims were in Arafat on that day. This distribution of cases reflects the density of pilgrims at Mina during those days.

A standard protocol of referral and management was followed (Khogali, 1982). Patients were referred to heat exhaustion wards, after being screened at reception and special heat illness forms were completed (Al-Marzoogi *et al.*, 1983). On arrival in the heat exhaustion ward the following steps were followed:

(i) Patient was put on a canvas bed, covered with white gauze and sprayed with tap water at room temperature.

(ii) Fans blew air from many directions.

(iii) Delirious or excited patients were sedated with valium 10 mg i.v.

(iv) 10 ml venous blood sample was sent for laboratory investigation of Na^+, K^+, glucose, urea and creatinine.

(v) A sample of urine was collected and examined.

(vi) Dehydration was corrected with 500 ml normal saline followed by dextrose, 5% in normal saline (0.5-1 ℓ).

(vii) Associated medical problems were handled according to the clinical findings and condition of the patient.

The regimen was followed in all heat exhaustion wards along the pilgrimage route. The only difficulty was collecting blood samples (due to the rush of work).

RESULTS

Age and Sex

All age groups were affected. The youngest age recorded was 6 yrs and the oldest 90 yrs. The age distribution of 347 patients is shown in Fig. 1. Forty six per cent of the patients were aged 40-59 yrs, while 29.9% were less than 40 yrs old. Seventy per cent (242) were male and 30% (105) female.

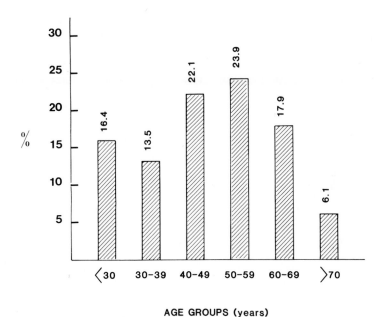

AGE GROUPS (years)

Fig. 1. Age distribution of 347 cases of heat exhaustion admitted to Mina General Hospital during the 1982 Hajj.

TABLE I.   *Nationality of 313 Heat Exhaustion Patients*

Nationality	Patients No.	%	Nationality	Patients No.	%
Egyptians	88	28.1	Indian	9	2.9
Yemeni	31	10.0	Indonesian	8	2.6
Saudi Arabian	23	7.3	Turkish	8	2.6
Algerian	21	6.7	Tunisian	7	2.2
Sudanese	17	5.4	Libyan	7	2.2
Syrian	16	5.1	Moroccan	7	2.2
Nigerian	15	4.8	Other	45	14.4
Pakistani	11	3.5			

Nationalities

Of the 347, the nationality was recorded for 313 patients, who represented 36 different nationalities mainly from the Middle-East and Asia.   Egyptians were the group most commonly affected, representing 28.1% of the total patients (Table I).

Body Temperature

The mean body temperature was 39.2°C.   Twenty seven per cent of the patients presented with body temperatures above 40°C and of these, 14 (4.4%) had temperatures above 41°C.

Clinical Presentation

Only 5% of the patients were semiconscious on admission, and the remaining 95% were conscious.
The main presenting symptoms were fever, weaknesses, giddiness and fatigue (Table II).

Skin

In 35.5% of patients there was hot skin on admission, whereas 24.5% had hot dry skin.   Skin flush was observed in 20.5% of the patients.   The others had hot wet skin.

TABLE II. *Presenting Symptoms of Heat Exhaustion Patients*

Symptoms	No. of patients	%
Fever	150	43.2
Weakness	70	20.2
Giddiness	51	14.7
Fatigue	46	13.3
Headache	21	6.0
Nausea and Vomiting	20	5.8
Dry mouth and tongue	15	4.3
Diarrhoea	8	2.3
Fainting	6	1.7
Muscle cramp	5	1.5

Pulse Rate and Blood Pressure

Normal pulse rate was recorded in 42% of the patients while the others had tachycardia. Of the latter group 10% had a pulse rate more than 120 per minute which returned to normal on cooling.

The mean B.P. was 128/78 Torr. The majority of patients (92.7%) were normotensive while 4.6% were hypertensive (highest B.P. = 200/110 Torr) and only 2.7% were hypotensive (lowest B.P. = 80/50 Torr).

Management of Patients

All patients were managed in well-ventilated aircondition-ed heat exhaustion wards at the reception where they received the above-mentioned regimen of treatment. Ninety three per cent of the patients received 0.5-1 ℓ of normal saline i.v. and the rate of infusion was adjusted in accordance with the general condition of the patient. Only 78% of the patients needed spraying and fanning, the other 22% improved on i.v. fluids and a stay in the airconditioned wards. In addition to the normal saline, 45% of the patients were given 0.5-1 ℓ of dextrose 5%; 15.6% received paracetamol tablets to relieve headache.

Of the 347 cases only 7 (2%) needed admission to the hospital. Reasons for admission were non-responsiveness to cooling within 2 h (2 patients), psychiatric manifestations (2 patients), bronchopneumonia (2 patients) and one patient was admitted for social reasons. All other patients (98%) recovered completely and were discharged within 2 h. The mean duration of stay at the heat exhaustion ward was 50 min. Forty two per cent of the patients recovered during 30 min, and only 14.5% needed more than 90 min stay.

DISCUSSION

Although pure forms of heat exhaustion rarely occur, one may divide heat exhaustion into that associated with predominant water-depletion and that associated with predominant salt-depletion (Knochel, 1974). The symptoms of water-depletion heat exhaustion include marked thirst, fatigue and weakness, discomfort and impaired judgement. In severe cases patients demonstrate delirium, hyperthermia and coma. On physical examination the body temperature is almost invariably elevated. On the other hand the symptoms of heat exhaustion due to salt-depletion consist of profound weakness, fatigue, severe frontal headache, giddiness, anorexia, nausea, vomiting, diarrhoea and skeletal muscle cramps. Hypotension and tachycardia are prominent findings. The body temperature is usually normal or subnormal.

The clinical picture presented by our patients was that of mixed heat exhaustion, although the high prevalence of pyrexia and thirst favour diagnosis of heat exhaustion due to water-depletion. Because laboratory investigations were not routinely done, due to inadequate staff, it was difficult to differentiate between the two categories of heat exhaustion. Thus management applied was that for mixed heat exhaustion.

The organizational set up for screening and early detection of heat illness during the pilgrimage was described by Al-Marzoogi *et al.* (1983). The early detection of cases of heat exhaustion ensured that the cases referred to heat exhaustion wards were in the early stages. Very few severe cases were seen; the severe forms might have progressed to the heat stroke stage and been treated separately at the heat stroke treatment centres.

In conclusion, the early detection of heat exhaustion cases is of paramount importance to conditions prevailing during the pilgrimage. As discussed by Lind (1983), heat exhaustion may be part of the one series of events leading to heat stroke. Hundreds of cases, which otherwise, might have developed into severe forms of heat exhaustion or into potentially lethal heat stroke, are treated within hours with complete recovery. It is our belief that with more active preventive measures, a large proportion of these cases can be avoided.

Acknowledgement

Our thanks to the Minister of Health of Saudi Arabia and to all the medical staff in the Western Region, specially those at Mina General Hospital - without their dedication such results could never have been achieved.

REFERENCES

Al-Marzoogi, A., Khogali, M. & El-Ergesus, A. (1983). Organizational set up for detecting, screening, treatment and follow-up of heat disorders. *In* "Heat Stroke and Temperature Regulation" (M. Khogali & J.R.S. Hales, eds) pp. 31-39. Academic Press, Sydney.

Khogali, M. (1982). Heat disorders with special reference to Makkah Pilgrimage. Monograph, Ministry of Health, Saudi Arabia.

Knochel, J.P. (1974). Environmental heat illness. *Arch. Intern. Med. 133*, 841-864.

Lind, A.R. (1983). Pathophysiology of heat exhaustion and heat stroke. *In* "Heat Stroke and Temperature Regulation" (M. Khogali & J.R.S. Hales, eds), pp. 179-188. Academic Press, Sydney.

16
Pathophysiology of Heat Exhaustion and Heat Stroke

Alexander R. Lind

Department of Physiology
St. Louis University School of Medicine
St. Louis, Missouri

It is argued that heat exhaustion and heat stroke simply represent different degrees of severity on a continuum of disordered thermoregulation when the total heat load exceeds the system's capacity to dissipate heat and body temperature rises to high levels. Factors which may precipitate either disorder or which may occur as complications afterwards are considered.

The pathophysiology of heat stroke remains elusive and there are many confusing and even conflicting observations in the literature (e.g., cf. Leithead & Lind, 1964). There are many reasons for that, for example: 1) the experimental induction of heat stroke in man is not feasible on ethical grounds; 2) data are therefore available only from victims after the onset of the disorder, where the time before admission to hospital and the start of treatment is highly variable; the longer that time is, the more severe are the consequences and the poorer is the prognosis. Furthermore, there is clear evidence that laboratory data vary with time even after admission to hospital; 3) there is a lack of standardization in laboratory measurements to be made so that there are no systematic, controlled data available from all studies and that contributes further to the confusion in the literature. Again the timing of obtaining laboratory data varies because the urgency of treatment must take precendence, and differences in timing doubtless affects the findings; 4) there may be differences in the prodromal events depending on the combinations of metabolic and environmental heat loads, as well as the variations of contributory factors such as the state of hydration, the level of acclimatization, intercurrent illness and so on, many of which cannot be established clearly in assessing the circumstances that existed in the prodromal phase.

The same problems apply, in general, to the analysis of heat exhaustion in field conditions. Laboratory studies of heat tolerance have been frequent, both for men at rest and at work in hot climates. Some of the subjects involved reach a point of heat exhaustion at or before the time they reach the arbitrary limits of core temperature and heart rate that have been taken to be ethically permissible in such investigations; those limits, while variable from study to study, are reasonably comparable and emerged from early experience of heat tolerance studies. It is important to note that while some men reach their limits of tolerance (heat exhaustion) within the predetermined limits of rectal temperature in those experiments, many do not and the data make it clear that there are always wide individual variations in the responses. The environmental conditions in which young, reasonably fit men at rest and at work can establish a steady-state of thermoregulation emerge with reasonable consistency from a variety of studies of heat tolerance for both acclimatized and unacclimatized men (cf. Lind, 1977). When the combination of metabolic and environmental heat loads exceed such levels, a steady-state cannot occur and the body temperature continues to rise until a heat disorder intervenes.

There are some obvious similarities in the description of the prodromal symptoms for both heat exhaustion and heat stroke (see below). Those similarities suggest that there is a common etiology for the two disorders and that the occurrence of either disorder represents two points on a continuum of the inability of the thermoregulatory system to establish a steady-state of core temperature. The allocation of one or the other disorder to the victims, may simply reflect individual variation in the ability to tolerate high core temperatures. Obviously, many other factors may have to be taken into account, such as the motivation of those exposed to such severe stress (Taylor & Lyman, 1972), along with the many factors that influence tolerance to heat as discussed by Robertshaw (1983). It is not unreasonable, however, to postulate that heat exhaustion may be regarded as an incipient heat stroke. If so, then it follows that those who are more tolerant of high core temperatures may provide the bulk of the pool of victims of heat stroke while their less-tolerant fellows more readily succumb to heat exhaustion.

It is worthwhile to consider descriptions of each disorder. Charles Leithead described the prodromal events in heat stroke as follows (Leithead & Lind, 1964): "Prodromal symptoms may be experienced for a matter of minutes to an hour or two before consciousness is altered or lost .... Patients however have complained (in retrospect) of headache, dizziness, numbness or drowsiness while restlessness, purposeless or uncoordinated movements, aggressiveness, mania, suicidal tendencies and mental confusion have been observed by their companions ..... In some cases, there is a clear account of several days' ill-health before the onset of coma and hyperpyrexia. Symptoms which have characterized the early phase in such cases include weakness, giddiness, thirst, anorexia, nausea, vomiting, diarrhoea, and muscle cramps.".

The reports of many observers during experiments on heat tolerance share common features regarding the subject's condition at or before the point of tolerance; such symptoms may also be sudden in onset and may result in syncope. Those comments were summarized (Lind, 1977): "First, clinical manifestations are commonly accepted as evidence by which to terminate the exposure. Headache, nausea or vomiting often precede heat exhaustion at which time high core temperatures and high heart rates are common. Some circumoral or facial pallor may occur, often associated with trembling of the legs and uncoordinated movements during work. There may be marked changes in personality; restlessness is common, and the subjects may become remote and morose, or euphoric and belligerent. The subjects experience lassitude and general subjective discomfort.".

Critical Tissue Temperature and Associated Phenomena

There have been many experiments to enquire into the ability of men to tolerate heat exposure, with or without accompanying exercise. A number of criteria have been proposed to terminate such exposures to heat. Experience from early studies has led most experimenters to set limits on the rectal temperature of 39–39.5°C or heart rates 160–180 beats/min as signals to withdraw men from the heat. Wyndham & his colleages (1968) have usually set a limit of 40°C for rectal temperature.

The development of the symptoms described above, have sometimes been related to the rate of rise of rectal temperature, rather than to its absolute level. For example, the subjects of Eichna *et al.* (1945) who complained of such

symptoms were exhorted to continue work. Those who were able
to do so felt and looked better towards the end of the 4-hour
exposure than they did in the first hour. But from Eichna's
data, as well as from others, there is good evidence that
some Caucasian subjects cannot continue to work in the heat
when their rectal temperature exceeds 39°C; indeed, in one
experiment, half of Eichna's subjects, despite exhortations,
reached their limit of tolerance when their rectal temperatures were between 38.5 and 39°C. It is not clear why
Wyndham's (1968) demonstration that Bantu rarely succumb
before their rectal temperatures reach 40°C is due to a
racial factor, to a higher level of acclimatization or to
greater motivation.

One experiment is of importance to us here. Wyndham *et
al.* (1968) exposed groups of 10 highly-acclimatized Bantu to
a variety of different nearly saturated climates, with wet-
bulb temperatures ranging from 35.5°C to 40°C. The subjects
were essentially at rest and one criterion for removing them
from the heat was the achievement of a rectal temperature of
40°C. In the hottest climate, that goal was achieved in an
average of just over 1.5 hours. In the coolest climate, all
10 men survived for 48 hours; during most of that time their
rectal temperatures ranged from about 39 to 40°C (occasionally and briefly exceeding 40°C). These results are important
to us because they show that prolonged hyperpyrexia does not,
of itself, lead to heat stroke, although any attendant dehydration would result in higher core temperatures and the
onset of frank heat disorder. It seems reasonable to conclude,
however, that for highly acclimatized subjects, heat stroke is
unlikely to occur at a rectal temperature of less than 40°C.
Most of Wyndham's subjects exhibited the commonly described
changes in personality and behaviour once their rectal temperatures reached 39-39.5°C, some showing periods of aggressiveness alternating with apathy, some being difficult to arouse,
some exhibiting bouts of weeping, etc. Therefore, although
they appeared to be stable in thermoregulatory terms they were
certainly not stable in emotional terms. Such observations
have led to the suggestion that hyperpyrexia is associated
with cerebral hypoxia. Shibolet *et al.* (1976) reject that
possibility on the basis of experiments on anaesthetized dogs
in which there was no evidence of changes in cerebral blood
flow or oxygen or glucose consumption at rectal temperatures
up to 42°C although differences did occur at higher rectal
temperatures. Cerebral hypoxia in hyperthermia has not been
examined in humans, but certainly the changes in measurements
in studies of an experimental-psychological nature, along with
the commonly found changes in personality and behaviour in

hyperpyrexic subjects would suggest that some hypoxia is present. The only alternative suggestion would be that hyperpyrexia by itself, without hypoxia, is responsible for the changes that have been described.

Prolonged hyperthermia in the presence of aerobic exercise with rectal temperatures of 39–40°C is not associated (Wyndham, 1973) with changes in serum enzymes to the levels found in heat stroke victims when they are admitted to the hospital. Increases in such enzymes are indicative of cellular damage. Similarly, long-distance runners who often raise their rectal temperatures to 40°C or above, show elevations of some serum enzymes but not nearly to the extent found in heat stroke cases. In heat stroke cases, the enzyme changes continue to rise for some 48 hours after the event, whereas in the hyperpyrexic subjects or marathon runners, they quickly decline.

Knochel (1974) makes a good case for the involvement of a deficit of potassium leading to thermoregulatory instability in men undergoing intensive physical activity in hot climates. Much of the potassium loss reported was in the sweat but there was also a loss in the urine, disproportionately high for men depleted of potassium. The hypothesis was that the intense physical work in the heat stimulated high renin and aldosterone production which could facilitate continued excretion of potassium by the kidney despite serious potassium depletion. While some aspects of Knochel's argument may have flaws, overall it has merit and is reminiscent of the pitressin-resistant polyuria found in some heat stroke cases in active military personnel.

Metabolic acidosis has also been suggested as a contributory factor in heat tolerance and in heat stroke. Wyndham (1973) summarized the existing data to show that a higher rectal temperature and a lower heat tolerance is associated with lower maximal aerobic capacity when men work at the same absolute rate; however, when they worked at the same proportions of their maximal aerobic capacities the rectal temperatures of all those men were the same. Clearly, if the men with lower aerobic capacities have to invoke anaerobic metabolism to do a given piece of work there is an inference that the resultant metabolic acidosis may be associated with lowered heat tolerance which may possibly contribute to the advent of heat stroke.

Little is known of acid-base, fluid or ionic changes in various tissues in man, although there are some fragments of evidence from animal studies, described briefly below.

A question that remains unanswered is what causes the very large individual variation in heat tolerance. It is not attributable simply to differences in criteria applied in different laboratories. It is found in all laboratory experiments and it is at its greatest in unacclimatized men. Whatever criterion is set, there are some subjects who tolerate hyperpyrexia in laboratory studies with no obvious untoward effects, long after their frailer colleagues have succumbed to heat exhaustion or who have reached the required hyperpyrexic criterion. The variation in aerobic capacity of tolerant and intolerant subjects can account for about half the variation seen and, so far, there is no evidence to account for the remainder.

Evidence from Animal Experiments on Heat Stroke

1. Tissue Destruction. It has been shown (e.g. Linke *et al.*, 1972) that a variety of animal tissues were destroyed within an hour at temperatures of 44-46°C. Exposure for 2 hours reduced those values to 42-45°C, while others have shown that brain tissue, red blood cells and many other tissues are damaged by exposure at 42°C. Interestingly, the upper limit of temperature for the growth of cells in tissue culture is theoretically set at 40.6°C.

2. Whole Body Heating.

a. Dogs. Shapiro *et al.* (1973) exposed dogs to hot environments both at rest and while exercising. The rectal temperature was controlled in various groups of dogs to be within given limits. Whether at rest or at work, no deaths occurred in dogs whose rectal temperature did not exceed 43°C; the exposure was discontinued when that temperature was reached and the duration of exposure was a mean of 89 min (range 30 to 270 min which indicates a degree of individual variability similar to that found in man). It is therefore not clear whether those dogs which reached a rectal temperature of 43°C in 30 min would have developed heat stroke if that level of temperature had been continued. Half of the dogs died when exposed to work in the heat but only one third died when exposed to heat at rest, when the rectal temperature was 43-44°C, inferring that exercise has a part to play in the development of heat stroke. All dogs whose rectal temperature exceeded 44°C died. In all these cases there was a relationship between the dogs which died and the time spent over the given rectal temperatures, expressed as (degrees x

min over 43°C);   the greater that product was, the greater
was the number of deaths that occurred.   That finding
reflects the clinical experience that the longer the interval
before treatment, the poorer the chances of survival.

While Frankel & Cain (1966) considered that there was no
evidence of cerebral hypoxia as judged by gross metabolic
exchanges in anaesthetized dogs whose core temperature reached as high as 42°C, Spurr & Barlow (1970) found that tissue
electrolytes of anaesthetized dogs heated to 42.5±0.5°C for
an hour showed significant increases in sodium in the liver,
jejunum and brain and of potassium in heart and skeletal
muscle.

b. Rats.   A rather extensive series of experiments in
conscious rats (e.g. Hubbard, 1979) produced some different
and some similar results to those found in dogs.   The authors
exposed conscious, resting rats to high environmental stress
and at exhaustive exercise in one cool (5°C) and 3 warm
environments (20, 23 and 26°C).   The rat neither pants nor
sweats.   None of the animals which ran to exhaustion at 5°C
died.   In the groups of rats which ran to exhaustion in the
warmer climates, a rectal temperature at 40.4°C represented
the threshold above which death occurred in a dose response
curve of a sigmoid character, with an LD_{50} of 41.5°C.   A
relationship of time spent at rectal temperatures over 40.4°C
(Δ°C x time) showed that increasing rates of mortality were
established as the product of (Δ°C x time) increased as in
the dog experiments of Shapiro *et al.* (1973).   Also, these
experiments clearly established that the involvement of
metabolic heat load resulted in greater numbers of fatalities
with respect to the relationships (degrees x minutes   spent
over 40.4°C)than occurred in the passively heated animals.
The differences varied by nearly 50% at LD_{25} reducing to
about 25% at LD_{75}, so that the fatality curves converged as
the product Δ of temperature x time increased.   As has long
been appreciated in clinical terms, the loss of life from
heat stroke is heavily dependent on the rapidity of treatment;   that is, reduction of core temperature.   No special
attempts were made to cool these animals quickly.

c. Other animals.   As is reported elsewhere in this volume (Robertshaw, 1983;   Hales, 1983;   Khogali *et al.*, 1983)
it has been shown that animals which store heat and also
depend on counter-current respiratory heat exchanges to
separate sharply the rectal and brain temperatures are capable
of surviving rectal temperatures which would be lethal in man
or in other animals.   The Thomson's gazelle is an extreme

example (Taylor & Lyman, 1972) which has been reported to survive a rectal temperature of up to 46.5°C for periods up to 6 h; brain temperature was much lower, by more than 3°C, as the result of an efficient counter-current respiratory heat exchange system. These observations are of great importance in drawing attention to the differences in rectal and brain temperatures in panting animals compared to non-panting animals in which such differences either do not exist or are minimized. Such evidence explains why the critical rectal temperature of dogs (panting) and rats (non-panting) are so very different (43°C as opposed to about 40.4°C).

To summarize, the inference is that it is the temperature of the central nervous tissue that is critical to the occurrence of heat disorders.

Laboratory Data on Heat Stroke in Man

The data from different reports on heat stroke are quite variable, largely because there is no standard procedure that is followed, while the complexities of timing, of the circumstances leading to the disorder and of individual variation doubtless play a part. For example, in most reports, the bulk of the victims were not sweating on admission whereas in the cases reported by Shibolet *et al.*, the contrary was true. Similarly, circulatory shock is commonly reported in about one of every 6 cases (cf. Leithead & Lind, 1964) whereas in the 'severe' cases reported by Shibolet *et al.* (1976) the rate was about four times greater, but was also in the ratio of 1:6 in 'mild' cases.

A serious problem in the blood-clotting mechanism occurs in most cases of heat stroke (Shibolet *et al.*, 1976). It is not clear whether this occurs at the onset of the disorder or is a later development. Capillary fragility and local haemorrhages in many tissues are commonplace and the attendant problems have been discussed by Shibolet *et al.* (1976). The viability of cellular membranes and the integrity of cellular function has been questioned but there is no direct evidence, other than on post-mortem study, to support that contention, reasonable though it may sound. Changes in body-fluids, plasma constituents etc. have been described elsewhere as well as in other contributions to this volume. However, at best, they represent an incomplete understanding of the events *following* heat stroke and those findings do not necessarily have a precise relationship to the circumstances that lead up to the occurrence of the disorder. Clearly, a great deal more needs to be established to characterize the prodromal events if we are to understand heat stroke and to be able to treat it properly. It is clear that in the past, many opportunities have been missed in experimental studies of heat tolerance and in the therapeutic use of hyperpyrexia to examine, in detail, physiological changes that are important. It is to be hoped that such opportunities will not escape in the future.

REFERENCES

Eichna, L.W., Ashe, W.F., Bean, W.B. & Shelley, W.B. (1945). The upper limits of environmental heat and humidity tolerated by acclimatized men working in hot environments. *J. Ind. Hyg. Toxicol. 27*, 59–84.

Frankel, H.J.M. & Caine, S.M. (1966). Arterial and cerebral venous blood substrate concentrations during hyperthermia. *Am. J. Physiol. 210*, 1265–1268.

Hales, J.R.S. (1983). Circulatory consequences of hyperthermia: An animal model for studies of heat stroke. *In* "Heat Stroke and Temperature Regulation" (M. Khogali & J.R.S. Hales, eds), pp. 223–240. Academic Press, Sydney.

Hubbard, R.W. (1979). Effects of exercise in the heat on prediposition to heatstroke. *Med. Sci. Sports 11*, 66–71.

Khogali, M., Elkhatib, C., Attia, M., Mustafa, M.K.Y., Gumaa, K., Nasr El-Din, A. & Al-Adnani, M.S. (1983). Induced heat stroke: A model in sheep. *In* "Heat Stroke and Temperature Regulation" (M. Khogali & J.R.S. Hales, eds), pp. 253–261. Academic Press, Sydney.

Knochel, J.P. (1974). Environmental heat illness. *Arch. Intern. Med. 133*, 1750–1758.

Leithead, C.S. & Lind, A.R. (1964). "Heat Stress and Heat Disorders", pp. 213–218. F.A. Davis, Philadelphia, PA.

Lind, A.R. (1977). Human tolerance to hot climates. "Handbook of Physiology" Section 9, Reactions to Environmental Agents (D. Lee, ed.), pp. 93–109. Williams and Wilkins Company, Baltimore.

Linke, C., Elbawadi, A. & Netto, V. *et al.* (1972). Effect of marked hyperthermia upon the canine bladder. *J. Urol. 107*, 599–602.

Robertshaw, D. (1983). Contributing factors to heat stroke. *In* "Heat Stroke and Temperature Regulation" (M. Khogali & J.R.S. Hales, eds), pp. 13–29. Academic Press, Sydney.

Shapiro, Y., Rosenthal, T. & Sohar, E. (1973). Experimental heat stroke: A model in dogs. *Arch. Intern. Med.* *131*, 688-692.

Shibolet, S., Lancaster, M.C. & Danon, Y. (1976). Heat Stroke: A review. *Aviation, Space and Environmental Medicine 47*, 280-301.

Spurr, G.B. & Barlow, G. (1970). Tissue electrolytes in hyperthermic dogs. *J. Appl. Physiol. 28*, 13-17.

Taylor, C.R. & Lyman, C.P. (1972). Heat storage in running antelopes: Independence of brain and body temperatures. *Am. J. Physiol. 222*, 114-117.

Wyndham, C.H. (1973). The physiology of exercise under heat stress. *Ann. Rev. Physiol. 35*, 193-220.

Wyndham, C.H., Williams, C.G., Morrison, J.F., Heyns, A.J.A. & Siebert, J. (1968). Tolerance of very hot humid environments by highly acclimatized Bantu at rest. *Brit. J. Ind. Med. 25*, 22-39.

17
The Elderly and their Risk of Heat Illness

K. E. Cooper and W. L. Veale

Department of Medical Physiology
The University of Calgary
Calgary, Alberta, Canada

There is now ample evidence from observations in man and from animal experiments that the thermoregulatory system may be modified with advancing age. Sweating and peripheral blood flow may change in the elderly and there is evidence also that thermoregulatory responses to the cold may, in a number of individuals, be profoundly altered and render the elderly more at risk of hypothermia. Experiments on old animals as compared with young animals seem to indicate that responses in body temperature to intravenously administered pyrogens differ in the older group of animals in that the second peak of the biphasic fever seen in young animals is lost. Furthermore, there is evidence that in the older animals, thermoregulatory responses to drugs such as mono-amines administered into the brain, may differ from that in young animals.

These differences, along with the ills which the elderly suffer which restrict the ability of their hearts and cardio-vascular systems to respond to very hostile thermal environ-ments, may all be of great importance in determining the incidence of heat stroke among an elderly population. The role which prescription drugs could also play in rendering an elderly population at risk is a matter which needs consider-able study.

There is considerable evidence that some thermoregulatory functions and the pattern of fever can alter with advancing years. Some of this evidence derives from observations on

HEAT STROKE AND
TEMPERATURE REGULATION
ISBN 0 12 406180 X

189

man and some from animal experiments. An epidemiological
study, carried out in St. Louis, Mo., USA, showed a relative
rise in hyperthermia deaths among the aged as compared to the
younger population (Schuman, 1972). Modified febrile
responses to intravenous administration of endotoxin have
been demonstrated in elderly rabbits by Ferguson *et al.*
(1981). There is also much evidence that, in man, the
incidence of death from hypothermia increases in the elderly
(Taylor, 1964). There is also evidence that there may be a
bimodal distribution of body temperature, with one peak in
the region of 37°C and another close to 35°C, in the elderly
(Collacott, 1975). Fox *et al.* (1973) recorded unusually low
temperatures in 10% of the elderly patients which were
studied. Although Finch & Hayflick (1977) found no age-
related changes in body temperature, and certainly the environmental conditions surrounding each set of observations are
of great importance in determining the body temperatures
observed, they remain in the minority of those reporting. It
is now important to review the reports that thermally
determined illness, namely heat illness and hypothermia,
occur more frequently in the elderly, and then to examine the
efferent mechanisms or underlying pathological processes
which may contribute to the problem.

HEAT ILLNESS, ARE THE ELDERLY REALLY AT RISK?

Many studies of heat illness have left the impression
that heat illness is primarily a problem of young people.
This is the result of the studies being performed on military
or industrial personnel, among whom there are few represent-
atives of the elderly (over 65) age group. However, more
extensive investigations of civilian populations during heat
waves, in the USA, demonstrated that 80% of the patients with
resulting heat illness were above 40 years of age (Leithead &
Lind, 1964).
A report by Collins (1934) indicated that the "excess"
deaths occurring during the mid-western USA heat wave of 1934
were mostly among elderly persons. However most of these
deaths were in people having cardiovascular or respiratory
system impairments. Obesity, another factor in the causation
of heat illness (Schickele, 1947), tends to occur most frequently in middle aged and elderly people. No doubt there is
plenty of evidence to support a high incidence of heat stroke
in the elderly, but it now becomes necessary to investigate
the relative roles of thermoregulatory deficits in the

elderly and of their other disease processes. We should also decide whether all or some of the deficits are common to all old people, or whether some of them are confined to a small segment of this age group.

PERCEPTION OF THE THERMAL ENVIRONMENT

Horvath *et al*. (1955) compared the ability of a group of subjects of 52 to 76 years of age with a group of young subjects in respect of their ability to sense the cold. The young subjects all found the cold environment uncomfortable, whereas the older subjects did not in spite of the fact that their deep body temperatures had fallen. Another group of volunteers with ages ranging from 74 to 86 years had thermoregulatory behaviour appropriate to the thermal discomfort but some felt cold only at unusually low environmental temperatures. Again Collins *et al*. (1977) found a reduced ability in the old to discriminate between changing environmental temperatures as compared to young subjects. Care had to be taken to distinguish between true failure to discriminate accurately the environmental temperature changes, and thermal confusion. A similar reduced temperature discrimination occurs in the elderly in hot environments (Collins *et al*., 1977). Any deterioration in the ability of the elderly to perceive changes in the environmental temperature, or indeed in their own internal state, can predispose to inappropriate voluntary responses to those changes and thus to heat illness.

SHIVERING

A study of eight elderly people, who had recovered from episodes of hypothermia, showed that they had lost the ability to respond to cold exposure by shivering, whereas people of the same age who had not suffered hypothermia could shiver (MacMillan *et al*., 1967). Collins *et al*. (1977) found that only 4 of his 43 elderly subjects shivered in the cold. The time of cold exposure was short, but more of the subjects complained of discomfort and cold than were observed to shiver. Horvath *et al*. (1955) also found that a number of people, over 52 years of age, failed to shiver when naked in a room at 10°C. While there is evidence that shivering may

not occur in many elderly people, whether this is a problem
of the entire age group or only of a sub-population still
remains to be decided.

VASOMOTOR RESPONSES

A study of forearm blood flow in elderly subjects, of
mean age 47 years, was made by Hellon & Lind (1958), and
their forearm blood flow changes induced by sitting and working in the heat were compared with those of young men of
average age 20 years. The environment was one with wet and
dry bulb temperatures of 30.5 and 38°C respectively. The
forearm flows of the older men were higher before and after
exercise, and the elevated flows were related to the skin
circulation, primarily. This observation would suggest that
if there is a reduced capacity for increasing the cardiac
output in the elderly, the increased demand for blood flow in
the skin could increase the level of cardiovascular strain.
Another study (Cooper, 1970), was made of hand blood flow
responses to raising the deep body temperatures in patients
of 68 to 94 years of age. In the patient of 94 years of age,
the hand blood flow response to core warming was lower than
in a younger age group, but the responses of the remaining 14
patients were within the range of those found in young healthy
adults. Similarly, the hand vasodilatation induced by exposure of the trunk skin to radiant heat, while widely variable
between subjects, was in the same range as that observed in
young people.
It seems then that there is no deterioration of the skin
blood flow responses to heat in the elderly, and indeed in
some areas of the body they may be enhanced. The temperature
at which the vasodilatory reflexes occur may be altered in
some of the elderly. Johnson & Park (1973) found that some
older people evidenced reflex vasodilatation, in response to
application of radiant heat to the trunk, at unusually low
body temperatures. Collins *et al.* (1977) classified the
alterations in vasomotor tone in response to environmental
temperature changes as follows:

(a) normal vasoconstriction or vasodilatation on cooling
or heating;
(b) absence of vasoconstriction on cooling but vasodilatation on heating;
(c) neither type of vasomotor response.

There is also some evidence (Fleisch *et al.*, 1980) of alteration in the sensitivity of the smooth muscle of blood vessels to vasoactive substances in older animals.

CHANGES IN SWEATING

There seems to be definite evidence that the threshold for sweating, in terms of deep body temperature, is raised in those of advanced years (Collins *et al.*, 1977; Fennell & Moore, 1973). This is so during passive heating or heating combined with exercise. Foster *et al.* (1976) showed a decrease in sweat rate in a number of skin areas, particularly in the extremities. It appears that there is a reduction in the quantity of sweat secreted by each sweat gland rather than in the number of active glands, as age increased. Post-menopausal women appear to have more severe reductions in their ability to sweat than do old men. It is known (Bannister, 1960), that circulating pyrogens, in amounts which produce only minimal symptoms, can arrest the process of sweating. It would be interesting to know whether this type of inhibition of sweating is exaggerated in the elderly.

ACCLIMATIZATION AND METABOLISM

Basal metabolism is known to be relatively reduced in older people. However little is known about the possible contribution of non-shivering thermogenesis to heat production in the elderly as compared to the young adult. Similarly, apart from the statement that the elderly may have a reduced capacity for acclimatization to heat, there is little evidence to indicate the cause of such reduced acclimatization. It will be important to distinguish between true decrements in "normal" acclimatization processes and limiting of the level of exercise, or disorders of water balance due to disease processes.

CHANGES IN THERMOREGULATION OBSERVED IN LABORATORY ANIMALS

The concentration of catecholamines in the preoptic area of the brain of the rat alters with increasing age (Estes & Simpkins, 1980). There are differences in the actions of the putative thermoregulatory neurotransmitter dopamine in young and older female rats (Kilbey *et al.*, 1980). Ferguson *et al.* (1983) showed that old rabbits get smaller fevers in response to intravenous endotoxin than do young ones. In addition, there is a reduction or abolition of the second peak of fever in the old animals. The fever in the old animals is accompanied by a large rise in the plasma levels of adrenaline. It appears that the older animals have the drive to elevate their temperatures at the time of the expected second fever peak, but cannot do so. The whole question of the function of the brain neurotransmitters in relation to alterations in thermoregulatory function in the old animals, and indeed in man, is wide open for study.

THE ELDERLY, DISEASE AND HEAT ILLNESS

Many disease states will interfere with the ability of the body to regulate its temperature, either in the cold or the heat, or in both. Where there is a demand for increased perfusion of the peripheral tissues, such as is needed in activating the heat loss by peripheral vasodilatation, or by the need to supply fluid to the sweat glands, diseases which severely restrict the cardiac output will predispose to failure of the thermoregulatory mechanism. Renal disorders which interfere with the body fluid balance will impair the ability to sweat. If there is a disorder of the hypothalamus or its efferent pathways, temperature regulation may fail. There may be minute lesions of the hypothalamus and brainstem, the only consequence of which is to alter the pattern of thermoregulation. Obesity, a common problem in many of the elderly, particularly the affluent elderly, may increase insulation, increase the strain on the cardiac capacity, and impair acclimatization. So, many problems, and some of the vices which go with advancing years, tend to put the older folk particularly at risk of suffering heat illness in hot climates, especially if they are not acclimatized or if they subject themselves to levels of exercise, or to dietary restraints to which they are not accustomed.

We need more studies of thermoregulation in the heat, in the elderly, and indeed some definitive studies of such things as resting body temperatures of older people and the circadian rhythms of body temperature. There is further need for studies of metabolic, circulatory and thermal perception in old people in order properly to design devices and regimens which will protect them from overheating during such thermal stresses as they experience in their pilgrimage to the hot holy places.

REFERENCES

Bannister, R.G. (1960). Anhidrosis following intravenous bacterial pyrogen. *Lancet ii*, 118–122.

Collacott, R.A. (1975). Screening for hypothermia in the Orkney. *J. Roy. Coll. Gen. Pract. 25*, 647–651.

Collins, S.D. (1934). *Publ. Hlth. Repts. (U.S.A.). 49*, 1015.

Collins, K.J., Dore, C., Exton-Smith, A.N., Fox, R.H., McDonald, I.C. & Woodward, P.M. (1977). Accidental hypothermia and impaired temperature homeostasis in elderly. *Brit. Med. J. 1*, 353–356.

Cooper, K.E. (1970). Studies of the human central warm receptor. *In* "Physiological and Behavioral Temperature Regulation" (J.D. Hardy, A.P. Gagge & J.A.J. Stolwijk, eds), pp. 224–230. Charles C. Thomas, Springfield, Ill., USA.

Estes, K.S. & Simpkins, J.W. (1980). Age related alterations in catecholamine concentrations in discrete preoptic area and hypothalamic regions in the male rat. *Brain Res. 194*, 556–560.

Fennell, W.H. & Moore, R.E. (1963). Responses of aged men to passive heating. *J. Physiol. (Lond.) 231*, 118–119.

Ferguson, A.V., Veale, W.L. & Cooper, K.E. (1981). Age related changes in the febrile response of the New Zealand white rabbit to endotoxin. *Can. J. Physiol. Pharmacol. 59*, 613–614.

Ferguson, A.V., Veale, W.L. & Cooper, K.E. (1983). Effector Mechanisms: their role in the age-related reduction in endotoxin fever in the rabbit. *Am. J. Physiol.* (in press).

Finch, C.E. & Hayflick, I. (1977). *In* "Handbook of the Biology of Ageing", pp. 159–188. Van Nostrand, New York.

Fleisch, J.H. (1980). Age related changes in the sensitivity of blood vessels to drugs. *Pharmac. Ther. 8*, 477–487.

Foster, K.G., Ellis, E.P., Dore, C., Exton-Smith, A.N. & Weiner, J.S. (1976). Sweat responses in the aged. *Age Ageing 5*, 41–101.

Fox, R.H., McGibbon, R., Davies, L. & Woodward, P.M. (1973). Problem of the old and the cold. *Brit. Med. J. 1*, 21–24.

Hellon, R.F. & Lind, A.R. (1958). The influence of age on peripheral vasodilatation in a hot environment. *J. Physiol. (Lond.) 141*, 262–272.

Horvath, S.M., Radcliffe, C.E., Hatt, B.K. & Spurr, G.B. (1955). Metabolic responses of old people to a cold environment. *J. Appl. Physiol. 8*, 145–148.

Johnson, R.H. & Park, D.M. (1973). Intermittent hypothermia. *J. Neurol. Neurosurg. Psychiat. 36*, 411–416.

Kilbey, M.M., Ellinwood, E.H. Jr., Gonzalez, L.P. & Cooper, R.L. (1980). Differential dopaminergic function in young and old female rats as measured by three behaviors. *Comm. in Psychopharm. 4*, 1–9.

Leithead, C.S. & Lind, A.R. (1964). "Heat Stress and Heat Disorders". Cassell, London.

MacMillan, A.L., Corbett, J.L., Johnson, R.H., Smith, A.C., Spalding, J.M.K. & Wollner, L. (1967). Temperature regulation in survivors of accidental hypothermia of the elderly. *Lancet ii*, 165–169.

Schickele, E. (1947). Environment and fatal heat stroke. *Milit. Surg. 100*, 235–256.

Schuman, S.H. (1972). Patterns of urban heat wave deaths and implications for prevention: Data from New York and St. Louis during July 1966. *Environ. Res. 5*, 59–75.

Taylor, G. (1964). The problem of hypothermia in the elderly. *Practitioner 193*, 762–767.

18
Drug-Induced Changes in the Thermoregulatory System

Peter Lomax

Department of Pharmacology
School of Medicine and the
Brain Research Institute
University of California
Los Angeles, California

Studies on the interaction of pharmacological agents and body temperature in man have been mainly related to induced toxicity or drug interactions. The effects of adverse environmental conditions, e.g. ambient temperatures below or above the thermoneutral range, on drug-induced changes, have been investigated in several sub-human species but studies of the responses in man under these conditions are far fewer. However, even although specific data may be lacking, one can predict the likely human responses from the results of animal research provided the basic underlying factors are elucidated. Such predictions require knowledge of normal thermoregulatory control mechanisms, understanding of the response of the system to changes in the environmental thermal load, elucidation of the basic mechanisms of the drug effect and its precise site of action on the thermoregulatory system. A further confounding factor is the underlying condition for which the therapeutic agent is being administered; cardiovascular and endocrine disorders are prominent examples in relation to heat stroke. Agents which may induce hyperthermia and predispose to heat stroke at high ambient temperatures are reviewed.

Descriptions of heat stroke and its clinical manifestations have been recorded since ancient times and of the several disorders which may result from exposure to hot environments it is the most serious, if not most common. Many predisposing factors have been identified including severe exercise, cardiovascular disease, advancing age and the use of alcohol (Ellis, 1972; Kilbourne *et al.*, 1982). An estimated 1,265 heat-related deaths may have occurred in the United States during the summer of 1980 (Impact assessment: US social and economic effects of the great 1980 heat wave and drought. Environ. Data Info. Service Nat. Ocean Atmosph. Admin. Sept. 7, 1980, p.12).

There have been many studies of the risk factors amongst specific groups of individuals such as military personnel (Stallones *et al.*, 1957) and athletes (Rose *et al.*, 1980). However, in the general population, in which many of the victims are elderly, the antecedants of heat stroke differ substantially from those in young athletes and military recruits. Kilbourne and his colleagues (Kilbourne *et al.*, 1982) conducted a case control study of such risk factors amongst heat stroke victims in St. Louis and Kansas City during the heatwave of July and August 1980. Data were gathered from 156 persons with heat stroke, of whom 73 died, and were compared with 462 matched controls. Using sophisti-cated statistical analysis 10 specific risk factors were identified. Amongst these, alcoholism and the use of certain drugs were associated with fatal heat stroke. The therapeuti agents implicated included phenothiazines, butyrophenones, thioxanthenes, antihistamines, tricyclic antidepressants, 'pure' anticholinergics, sedative hypnotics, sympathomimetics phosphodiesterase inhibitors, thyroid agonists and diuretics. Frequently the victims were taking more than one drug, e.g. subjects taking antiparkinson anticholinergics were also using one of the major tranquilizers. The majority of clinical reports of severe or fatal hyperpyrexial reactions associated with drug use have involved the treatment of psychiatric patients. Such reactions have occurred not only from an overdose of the drug, but also at conventional dose levels and particularly during treatment with two or more compounds. The use of psychotropic agents by psychiatrists, as well as physicians and general practitioners, is increas-ing, and they are more and more frequently being used in combination. This, together with the increased risk of self poisoning in depressed patients, makes it likely that the incidence of drug-induced hyperpyrexia will increase. In such cases hyperpyrexia and heat stroke may occur not only during exposure to high environmental temperatures but also under less extreme conditions, and even at normal ambient temperatures.

MECHANISMS OF DRUG-INDUCED HYPERPYREXIA

The effects of drugs on temperature regulation have been studied extensively in animals but there have been considerably fewer studies in man (review: Lomax & Schönbaum, 1979). There is far less information relating to responses during exposure to high ambient temperature; nor have the important questions of changes in drug distribution and metabolism during hyperpyrexia seemingly been addressed at all.

Reports of clinical disorders of thermoregulation related to drug administration, particularly during heat waves, are to be found in the literature. From a knowledge of the mechanism of action of the drug in question, and an understanding of the normal thermoregulatory system, it is possible to predict with some confidence the underlying pathophysiological mechanisms responsible for the pyrexia. Conversely, on the basis of the same fundamental information, one might be able to forecast the effect of a given compound on body temperature under conditions of thermal stress. Such considerations may be valid in reference to the normal therapeutic effects of the drug and to its known toxic conditions but serious clinical disturbances can occur under more severe thermal stress - high or low environmental temperatures.

SPECIFIC DRUGS

A. *Phenothiazines*

The phenothiazines are amongst the most widely used drugs in current medical practice, primarily for the treatment of psychiatric patients. Disorders of body temperature are a recognized risk of therapy with phenothiazines, especially in elderly patients; both hypothermia (Glinoer *et al.*, 1973; Exton-Smith, 1972; Laughnasse, 1968) and hyperthermia (Ayd, 1956; Ext, 1958; Zelman & Guillan, 1970) may occur with the ambient temperature being the determining variable. Body temperatures as high as 42°C have been reported in adults treated with phenothiazines. Although, in some of these cases, hyperthermia followed overdosage, or very high therapeutic doses, in others the condition developed following normal therapeutic regimens. Common to all cases of hyperpyrexia was exposure to high environmental temperatures and/or physical exercise. Flushing and dry skin were characteristic

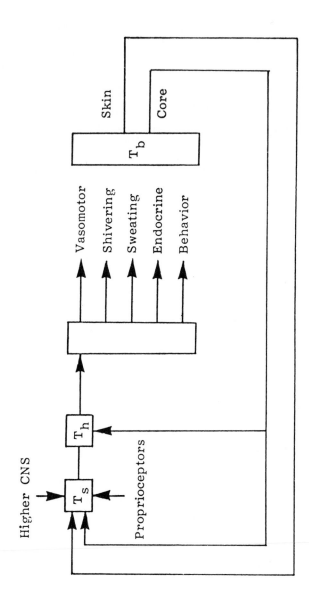

Fig. 1. Model of the thermoregulatory system based on proportional control theory with a variable set temperature (T_s). Inputs from thermosensors in the superficial (skin) and deep (core) tissues, from proprioceptors and from higher levels of the central nervous system (CNS) determine the thermoregulatory set-point. The set-point and hypothalamic thermosensors (T_h) are compared and according to the direction and magnitude of the offset the appropriate effector systems are activated to prevent any change in deep body temperature in the face of the altered environmental conditions.

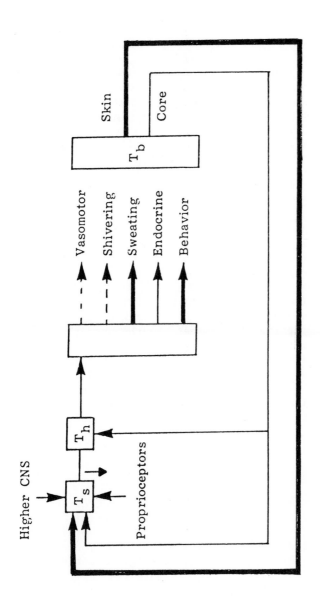

Fig. 2. Response of the thermoregulatory system to exposure to a high environmental temperature. Signals, from stimulation of warm thermal receptors in the skin, to the hypothalamus lower the set-point. T_s is now less than T_h so heat loss mechanisms are activated (sweating, appropriate behavioural responses) and vasomotor tone to the cutaneous vessels is decreased causing increased skin perfusion and heat loss from radiation. Thus, any potential rise in core temperature will be prevented providing the external heat load does not exceed the capacity of the system or the system is not defective (e.g. sweating is impaired by dehydration).

and sometimes the patients developed convulsions. The risk of hyperthermia is greater with the longer acting compounds such as fluphenazine enanthate (Lemoine, 1973).

Since the primary therapeutic effect of the phenothiazines results from their central actions it seems to have been generally assumed that the thermoregulatory disorders were due to disruption of the central nervous system regulatory mechanisms. Generally, animal experiments have not substantiated this belief. The quaternary derivative of chlorpromazine, which does not traverse the blood-brain barrier, causes changes in body temperature in the rat comparable to that seen with the parent molecule (Kirkpatrick & Lomax, 1971). In a cold, or thermoneutral, environment the temperature falls and at ambient temperatures above the thermoneutral zone (>30°C) hyperthermia occurs. Injection of chlorpromazine, or the N-methyl derivative, directly into the preoptic/anterior hypothalamic nuclei (site of the thermoregulatory centres) causes a rise in body temperature in rats maintained at an air temperature of 18°C (Kirkpatrick & Lomax, 1971). This effect of centrally injected chlorpromazine is similar to that of atropine (Kirkpatrick & Lomax, 1967), and chlorpromazine has marked cholinolytic properties at peripheral sites such as the guinea pig ileum and superior cervical ganglion (Lomax & Kirkpatrick, 1974).

It seems unlikely that the pyrexia in man could be due to a resetting of the central thermostats by the phenothiazines since body temperature is not perturbed significantly under thermoneutral conditions. Under conditions of heat stress, particularly when associated with exercise, heat dissipating mechanisms seem to be deranged and the drugs may inhibit sweating by selectively blocking the postganglionic nerve supply to eccrine glands by virtue of their antimuscarinic activity.

B.   *Tricyclic Antidepressants*

The effects of the tricyclic antidepressant drugs on body temperature have been reviewed recently by Garattini & Jore (1979). Generally hypothermia is seen and the magnitude of the effect of the individual compounds is a function of their antipsychotic phenothiazine-like pharmacological activity. More important, in the present context, are the interactions between tricyclics and other agents which affect body temperature, such as reserpine, amphetamine, noradrenergic and

cholinomimetic drugs. These interactions have been widely
used to evaluate the antidepressant effects of these drugs in
laboratory animals (Garattini & Jore, 1979; Loskota &
Schönbaum, 1979).

Hyperpyrexia and heat stroke have been reported after
overdose with a combination of a monoamine oxidase inhibitor
(tranylcypromine), a β-adrenergic blocking agent (propranolol)
and a benzodiazepine (King *et al.*, 1979). On admission to
hospital this patient exhibited hot, dry skin, a rectal
temperature of 42°C, muscle rigidity and coma. Sweating was
completely absent. It seems likely that enhanced adrenergic
activity due to tranylcypromine played an important role in
this syndrome but whether this was a result of central or
peripheral actions is unclear. However, excess adrenergic
activity at peripheral nerves has been implicated in the
fulminant hyperthermia stress syndrome in swine and administration of phentolamine, an α-blocker, has been reported to
prevent, or reduce the severity, of the attacks (Williams &
Stubbs, 1977).

C. *Opiates*

The opiate alkaloids, especially morphine, have profound
effects on body temperature, but the responses vary considerably according to the dose and the particular species; in
some species the predominant response is a fall in temperature while in others hyperthermia occurs (see Burks &
Rosenfeld, 1979). With therapeutic doses there is no significant effect on core temperature in man although dilatation
of cutaneous vessels, especially of the face, neck and upper
thorax, giving rise to a local increase in skin temperature
and sweating is seen, partly due to histamine release.

More marked effects are seen during opiate withdrawal in
dependent individuals. Within 8-16 h of withdrawal profuse
sweating and piloerection are seen; by 36 h muscle fascicu-
lation, twitches, muscle cramps and a rise in body temperature
occur.

A much more serious hyperpyrexial reaction was first
described by Mitchell (1955) following administration of
meperidine (pethidine) to individuals previously treated with
monoamine oxidase (MAO) inhibitors, including iproniazid,
phenelzine, pargyline and tranylcypromine (see Gessner, 1973).
A fatal toxic reaction, with hyperpyrexia, has been reported
following ingestion of a cough mixture containing dextromethorphan by a woman who had been receiving MAO inhibitor
therapy (Rivers & Horner, 1970). Such patients develop clonic

spasms, trismus, hypertension, tachycardia, cyanosis, coma
and severe hyperpyrexia which, unless vigorously treated can
lead to a fatal outcome. A similar syndrome can be induced in
rabbits and mice (Rivers & Horner, 1970). Experiments in
these animal models suggest that the toxicity of meperidine
after MAO inhibitors is the result of accumulation of serotonin in the brain (Rivers & Horner, 1970; Rogers & Thornton,
1969). Meperidine is unique among the narcotic analgesics in
inhibiting serotonin re-uptake by neurons (Carlsson &
Lindquist, 1969). Thus, the blockade of uptake and metabolism
leads to the accumulation of the indoleamine and this could
have a direct effect on the thermoregulatory centres leading
to a raising of the set-point in these species (see Jacob &
Girault, 1979).

D. *Methyldopa*

An elevation of body temperature may occur during the
first few weeks of treatment with methyldopa in patients with
hypertension. In some cases this may be due to haematological or haepatic disorders, but it may also result from
the increased level of circulating sympathomimetic amines
formed by decarboxylation of methyldopa and subsequent
cutaneous vasoconstriction. Such febrile reactions have been
reported in up to 5% of patients treated with the drug.
While such hyperthermia is of little consequence in temperate
zones, significant hyperpyrexia could develop during heat-
waves.

E. *Amphetamines*

The importance of the naturally occurring catecholamines
in temperature regulation is well documented even if not
completely understood. Compounds which have similar direct
actions on the tissue receptors, or affect the release or
metabolism of catecholamines, would be expected to have
similar thermoregulatory effects. Amongst such drugs the
amphetamines are of particular interest because of their
widespread, if questionable, use as appetite suppressants in
hyperkinesis in children and as central nervous system stimulants. Also, because of the central stimulant effect, these
are one of the most widespread group of drugs of abuse.
Hyperpyrexia and fatal heat stroke are relatively common
occurrences in acute amphetamine poisoning. In a detailed
review Sellers and his colleagues (1979) have described the

clinical syndrome in two of their own patients and in 28 reports in the literature. There were 11 deaths in these patients. Fatal or near fatal syndromes associated with hyperthermia were less common among chronic abusers of high doses of amphetamines but were more frequent in the relative neophyte taking accidental, or intentional overdoses (Ellinwood, 1975).

Exertion and high ambient temperatures are important factors in fatal episodes, and exposure to high altitudes may be a contributing factor. Concurrent administration of other drugs such as alcohol, barbiturates, MAO inhibitors and phenothiazines has been reported in fatal cases. The toxicity and pyrexial effects of amphetamines are increased in hyper-thyroid patients, or in individuals treated with thyroxine.

There have been many experimental studies to determine the relative importance of central and peripheral actions of amphetamines on temperature regulation; certainly both loci are important in the induced hyperpyrexia. Animal experi-ments have generally proved to be inconclusive with respect to effects on the thermoregulatory centres; in most species studied intracerebral injection of amphetamine leads to a fall in core temperature. The effects of amphetamine on dop-amine, norepinephrine and serotonin metabolism in the central nervous system fail to provide an explanation for the hyper-pyrexia seen in man. Similarities to the increased skeletal muscle heat production seen during exercise and shivering and in malignant hyperthermia, in which norepinephrine plays a mediating role in the thermogenic process occurring in muscle cells, suggests that activation of peripheral heat production is a major factor in the development of pyrexia. Thus, although amphetamine-associated hyperthermia is not too common, considering the widespread abuse of the drug, its clinical importance is clear and further research, especially under conditions of high ambient temperatures, is clearly needed.

F. Cannabinoids

In laboratory animals, at ambient temperatures within or below the thermoneutral range, tetrahydrocannabinol (THC) causes a fall in body temperature, while at high environmen-tal temperatures hyperthermia results (see Haavik & Hardman, 1979). Studies, using doses of marihuana in the range associated with smoking the drug ($\sim$15-20 mg THC), on body temperature in man in a normal or cool environment revealed no changes or decreases of 0.1-0.2°C. Adult humans are

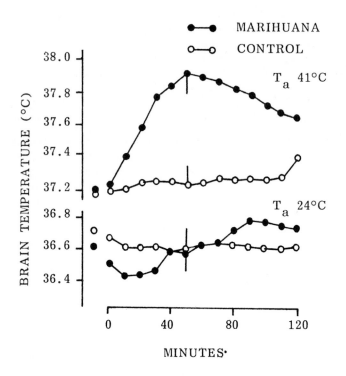

Fig. 3.   The effect of smoking a marihuana cigarette [containing THC (18 mg)] or a placebo cigarette on tympanic membrane temperature in groups of 6 adults at an ambient temperature (Tₐ) of 40±2°C and 12 adults at Tₐ 23±1°C.   Single mean values on left are those 20 min prior to smoking. Vertical bars represent 1 S.E.M.   Data from Jones et al. (1980), courtesy of Karger.

generally more resistant to disruptions of body temperature than are smaller animals, possibly due to the lesser ratio of body surface area to volume. Jones and his colleagues (1980) compared the effects of THC (18 mg) and placebo cigarettes on groups of healthy volunteers at ambient room temperatures of 23±1°C or 40±2°C. In subjects smoking THC at the high environmental temperature there was a significant rise in core (tympanic membrane) temperature ranging from 0.6–1.3°C. Skin (finger) temperature showed an increase which paralleled the core temperature. In 5 of 6 subjects smoking marihuana at the elevated room temperature sweating was suppressed after 20 min and the subjects reported feeling less hot.

These latter findings are consistent with an upward setting of the central thermostats induced by THC. These findings suggest that hyperpyrexia could readily occur with marihuana use at high ambient temperatures or when thermoregulatory capacity is already limited (Jones *et al.*, 1980). These results are illustrated in Fig. 3.

G. *Antimuscarinic Agents*

The belladonna alkaloids induce a rise in body temperature in man which is usually significant only after high doses or during exercise at high environmental temperatures. In infants and small children even moderate doses may induce "atropine fever". There is one case report of hyperpyrexia leading to death in a patient given atropine (0.2 mg) prior to a bronchoscopy (Chapman & Bean, 1956). Suppression of sweating is doubtless a considerable factor in the genesis of fever when the ambient temperature is high. However, the circulatory disturbances and cerebral irritation caused by injection of therapeutic doses of atropine (2 mg i.m.) into men working under hot dry conditions are more damaging than the inhibition of sweating produced by the drug (Cullumbine & Miles, 1956).

Atropine may exert a central effect on thermoregulation, especially in high doses. Direct injection of atropine into the thermoregulatory centres of experimental animals (Kirkpatrick & Lomax, 1967) caused a rise in core temperature; cholinomimetics and inhibitors of cholinesterase, conversely, lead to hypothermia in animals and man, which is a consequence of a lowering of the thermoregulatory set-point (Cox *et al.*, 1975). Thus, a rise in the set temperature may contribute to the hyperpyrexia when sweating is suppressed by antimuscarinic drugs.

H. *Alcohol and Barbiturates*

Alcoholism and moderate to high doses of ethanol and barbiturates are commonly quoted in the literature as risk factors for heat stroke (e.g. Kilbourne *et al.*, 1982). Case reports specifically implicating these drugs seem to be lacking; in fact both are more usually associated with cases of accidental hypothermia. Ethanol, in doses leading to blood levels in experimental animals which would be found in only moderate intoxication in humans, causes a downward setting of the central thermostats (Lomax *et al.*, 1980). In high doses,

barbiturates induce coma and depress all thermoregulatory
systems (including appropriate behaviour) so that the
individual is essentially poikilothermic and the body temperature will fall or rise depending on the external environment.
During exposure to high ambient temperatures even moderate
doses of these compounds may interfere with normal judgement
to the extent that protective measures are not taken and heat
stroke supervenes.

THE EFFECT OF TEMPERATURE ON THE ACTION OF DRUGS

Much has been published on temperature regulation and the
physiology and pharmacology of heat and cold. However, discussion of the specific question of changes in drug action by
temperature or climate in man are rare, and there are almost
no papers in the field of clinical pharmacology (Fuhrman,
1963), in spite of many studies in experimental animals (see
Weihe, 1973). Presumably physicians have not considered the
thermal environment as an important variable. Nor have
questions concerning changes in drug metabolism and distribution under differing environmental heat stresses been
addressed. In the case of drugs where the therapeutic dose
is close to that which will give rise to adverse (toxic)
effects (i.e. compounds with a low therapeutic index) small
changes in these parameters would be expected to have marked
clinical effects. Of particular interest in this respect are
agents used for the management of cardiovascular diseases and
hormone treatment.
In a consideration of the etiological factors and pathophysiology of heat stroke there is a clear need for basic
research in this important area. Once physicians, as well as
physiologists and pharmacologists, become aware of the implications of the thermal environment they will notice modifications of drug action in man and may be able to relate these
to the environmental thermal conditions.

HEAT ACCLIMATION

With exposure to high environmental temperatures for
relatively short periods ($\sim$12 days) significant acclimation
can be demonstrated in human subjects; in particular there
are changes in the rates of protein transfer between fluid
compartments such that haemodilution is accomplished more

rapidly (Senay, 1975). There do not appear to be any reports of the effect of acclimation on the response to drugs which induce hyperthermia in man. Nor are there any detailed studies of the question in experimental animals although it has been demonstrated that the lethal hypothermic effect of ethanol in rats exposed to low (5°C) ambient temperatures can be completely abolished by cold acclimation for as short as 7 days (Lomax & Lee, 1982). Clearly, the development of appropriate animal models for the study of heat acclimation and drug effects in hot environments is urgently needed.

REFERENCES

Ayd, F.J. (1956). Fatal hyperpyrexia during chlorpromazine therapy. *J. Clin, Exp. Psychopathol. 27*, 189-192.
Burks, T.F. & Rosenfeld, G.C. (1979). Narcotic analgesics. *In* "Body Temperature" (P. Lomax & E. Schönbaum, eds), pp. 531-549. Dekker, New York.
Carlsson, A. & Lindquist, M. (1969). Central and peripheral monoaminergic membrane pump blockade by some addictive analgesics and antihistamines. *J. Pharm. Pharmacol. 21*, 460-464.
Chapman, J. & Bean, W.B. (1956). Iatrogenic heat stroke. *J. Am. Med. Assocn 161*, 1375-1377.
Cox, B., Green, M.D. & Lomax, P. (1975). Behavioural thermo-regulation in the study of drugs affecting body temper-ature. *Pharmacol. Biochem. Behav. 3*, 1051-1054.
Cullumbine, H. & Miles, S. (1956). The effect of atropine sulphate on men exposed to warm environments. *Quart. J. Exp. Physiol. 41*, 162-179.
Ellinwood, E.H. (1975). Treatment of reactions to amphet-amine-type stimulants. *Curr. Psychiat. Ther. 15*, 163-169.
Ellis, F.P. (1972). Mortality from heat illness and heat-aggravated illness in the United States. *Environ. Res. 5*, 1-58.
Ext, H.J. (1958). Five cases of heat stroke observed in mentally ill patients treated with Pacatal during the hot weather spell. *N.Y. State J. Med. 58*, 1877-1891.
Exton-Smith, A.N. (1972). Phenothiazines in cold weather. *Brit. Med. J. 1*, 441.
Fuhrman, F.A. (1963). Modification of the action of drugs by heat. *Arid Zone Res. 22*, 223-238.
Garattini, S. & Jore, A. (1979). Tricyclic antidepressant drugs. *In* "Body Temperature" (P. Lomax & E. Schönbaum, eds), pp. 439-459. Dekker, New York.

Gessner, P.K. (1973). Body temperature correlates of the interaction between monoamineoxidase inhibitors and meperidine. *In* "The Pharmacology of Thermoregulation" (E. Schönbaum & P. Lomax, eds), pp. 473-481. Karger, Basel.

Glinoer, D., Ectors, M., Poulet, P., Thys, J.P. & Cornil, A. (1973). Etude clinique de 39 observations d'hypothermie accidentelle de l'adulte. *Acta Clin. Belg. 28*, 40-55.

Haavik, C.O. & Hardman, H.F. (1979). Cannabinoids. *In* "Body Temperature" (P. Lomax & E. Schönbaum, eds), pp. 499-529. Dekker, New York.

Jacob, J.C. & Girault, J-M. (1979). 5-Hydroxytryptamine. *In* "Body Temperature" (P. Lomax & E. Schönbaum, eds), pp. 183-230. Dekker, New York.

Jones, R.T., Maddock, R., Farrell, T.R. & Herning, R. (1980). Marijuana and human temperature regulation in a hot environment. *In* "Thermoregulatory Mechanisms and their Therapeutic Implications" (B. Cox, P. Lomax, A.S. Milton & E. Schönbaum, eds), pp. 62-64. Karger, Basel.

Kilbourne, E.M., Choi, K., Jones, S. & Thacker, S.B. (1982). Risk factors for heatstroke. *J. Am. Med. Assocn 247*, 3332-3336.

King, I., Barnett, P.S. & Kew, M.C. (1979). Drug-induced hyperpyrexia. *South African Med. J. 56*, 190-191.

Kirkpatrick, W.E. & Lomax, P. (1967). The effect of atropine on the body temperature of the rat following systemic and intracerebral injection. *Life Sci. 6*, 2273-2278.

Kirkpatrick, W.E. & Lomax, P. (1971). Temperature changes induced by chlorpromazine and N-methyl chlorpromazine in the rat. *Neuropharmacology 10*, 61-66.

Laughnasse, T. (1968). Hypothermia in a young adult. *Lancet ii*, 455-465.

Lemoine, P. (1973). Psychotropes et régulation thermique. I. Les psycholegtiques. *Eur. J. Toxicol. 6*, 5-23.

Lomax, P., Bajorek, J.G., Chesarek, W.A. & Chaffee, R.R.J. (1980). Ethanol-induced hypothermia in the rat. *Pharmacology 21*, 288-294.

Lomax, P. & Kirkpatrick, W.E. (1974). Interactions of chlorpromazine with monoamines in the hypothalamic thermoregulatory centers. *In* "The Phenothiazines and Structurally Related Drugs" (I.S. Forrest, C.J. Carr & E. Usdin, eds), pp. 719-729. Raven, New York.

Lomax, P. & Lee, R.J. (1982). Cold acclimation and resistance to ethanol-induced hypothermia. *Europ. J. Pharmacol. 84*, 87-91.

Lomax, P. & Schönbaum, E. (eds) (1979). "Body Temperature: Regulation, Drug Effects and Therapeutic Implications". Dekker, New York.

Loskota, W.J. & Schönbaum, E. (1979). Reserpine. *In* "Body Temperature" (P. Lomax & E. Schönbaum, eds), pp. 427–437. Dekker, New York.

Mitchell, R.S. (1955). Fatal toxic encephalitis occurring during iproniazid therapy in pulmonary tuberculosis. *Ann. Intern. Med.* 42, 417–424.

Rivers, N. & Horner, B. (1970). Possible lethal reaction between Nardil and dextromethorphan. *Canad. Med. Ass. J.* 103, 85.

Rogers, K.J. & Thornton, J.A. (1969). The interaction between monoamine oxidase inhibitors and narcotic analgesics in mice. *Brit. J. Pharmacol.* 36, 470–480.

Rose, R.C., Hughes, D., Yarborough, D.R. *et al.* (1980). Heat injury among recreational runners. *South. Med. J.* 73, 1038–1040.

Sellers, E.M., Roy, M.L., Martin, P.R. & Sellers, E.A. (1979). Amphetamines. *In* "Body Temperature" (P. Lomax & E. Schönbaum, eds), pp. 461–498. Dekker, New York.

Senay, L.C. (1975). Plasma volumes and constituents of heat-exposed men before and after acclimatization. *J. Appl. Physiol.* 38, 570–574.

Stallones, R.A., Gould, R.L., Dodge, H.J. *et al.* (1957). An epidemological study of heat injury in army recruits. *Arch. Indust. Health* 15, 455–465.

Weihe, W.H. (1973). The effect of temperature on the action of drugs. *In* "The Pharmacology of Thermoregulation" (E. Schönbaum & P. Lomax, eds), pp. 155–169. Karger, Basel.

Williams, C.H. & Stubbs, D.H. (1977). Preliminary studies on the therapeutic efficacy of phentolamine in the fulminant hyperthermia-stress syndrome. *In* "Drugs, Biogenic Amines and Body Temperature" (K.E. Cooper, P. Lomax & E. Schönbaum, eds), pp. 233–234. Karger, Basel.

Zelman, S. & Guillan, R. (1970). Heat stroke in phenothiazine treated patients: a report of three fatalities. *Am. J. Psychiatry* 126, 1787–1790.

19
Face Fanning:
A Possible Way to Prevent or Cure Brain Hyperthermia

Michel Cabanac*

Département de Physiologie
Faculté de Médecine
Université Laval
Québec, P.Q., Canada

In mammals and birds the anatomy of the circulatory system of the head permits selective cooling of the brain during hyperthermia. A comparable system has been hypothesized to also exist in humans. The influence of this system, as a protector of the brain against hyperthermia, was studied in humans during dehydration and after exhausting muscular exercise.

During moderate dehydration oesophageal temperature (T_{es}) but not tympanic temperature (T_{tymp}) was elevated; sweat secretion was inhibited on the back of the subjects but not on their forehead.

During the recovery period after an exhausting race T_{tymp} decreased rapidly when the subject's face was fanned, but increased by 0.5°C and remained elevated when there was no fanning.

The results of both experiments confirm, indirectly, the existence of selective brain cooling in humans, and suggests a simple, cheap, and efficacious method for the prevention and therapy of brain hyperthermia, viz., face fanning.

Selective brain cooling has been shown to take place in many animal species during hyperthermia (reviews: Baker, 1982; Caputa, 1980). Humans also possess a selective brain

On leave from Université Claude Bernard, France.

cooling mechanism (Cabanac & Caputa, 1979a). The sweat secreted on the face evaporates and cools the blood in the capillary bed of the vasodilated skin. In hyperthermia, the blood of the face is collected to the *angularis occuli* veins and flows from face to brain and during hypothermia from brain to face (Caputa, Perrin & Cabanac, 1978). Such a mechanism was shown to operate during passive hyperthermia resulting from external heating, as well as active hyperthermia resulting from muscular exercise (Cabanac & Caputa, 1979b). It was of further interest to explore in the laboratory this mechanism under circumstances comparable to those of natural hyperthermia: dehydration (Cabanac, Caputa & Massonnet, unpublished) and prolonged muscular work (Germain, Jobin & Cabanac, 1983).

DEHYDRATION

Evaporation of water is the main avenue for heat loss when ambient temperature is above thermal neutrality. Since water storage within the human body is limited, prolonged sweating in response to elevated ambient temperature and/or muscular exercise, results in concentration of body fluids (Harrison, 1974; Senay, 1979). As a result of dehydration the overall rate of sweating decreases (Senay, 1979) and deep body temperature seems to be regulated at a higher level whether the subjects are at rest or at work (Strydom & Holdsworth, 1968; Nielsen, 1974; Nadel, Fortney & Wenger, 1980). The elevation of deep body temperature is proportional to the percentage of water deficit (Adolph *et al.*, 1947).

Moderate dehydration was produced in human subjects who refrained from ingesting any liquid for 24 to 36 h, sweated during short bouts of exercise on a cycle ergometer and spitted occasionally the day before the experiment. As a result the subjects experienced intense thirst and asthenia and their blood osmolality increased by 9.3 ± 1.4 m osm; their skin tended to keep the fold when pinched.

Three subjects served twice in dehydration and twice in control conditions of normal hydration. Dressed in bathing trunks and shoes, they sat on a cycle ergometer. Oesophageal temperature below the heart, and skin temperature on the back and forehead were recorded continuously. Every other minute, local sweating was measured alternatively on the forehead and on the scapula with a sudorimeter.

After a ten minute rest at 24°C, the subjects worked at a moderate intensity of 80 to 150 W for 20 min. They were then allowed to rest for 20 min.

Temperatures and Sweating Rates at Rest

Fig. 1 summarizes the results, and tables I and II show the statistical analysis. It can be seen that during normal hydration, oesophageal temperature (T_{es}) was not significantly different from the tympanic temperature (T_{tymp}) in control conditions. However, during dehydration, T_{es} was significantly (0.3°C) higher than T_{tymp}. Dehydration reduced sweating on the back from 19.4 g m^{-2} h^{-1} in control conditions to 7.5 g m^{-2} h^{-1}, i.e. reduced it 2.6 times. However, the forehead's sweat rate was the same during dehydration 26.4 g m^{-2} h^{-1}, and control conditions, 26.2 g m^{-2} h^{-1}.

The chronic elevation of deep body temperature during dehydration was therefore confirmed here. However, this increase was unchanged or even lowered by exercise. It is possible to explain this difference on the sole basis of sweat secretion. The above results show that evaporative heat loss was inhibited on the trunk but not on the face. Thus, selective brain cooling was maintained by sweat evaporation on the forehead and the face.

It may be hypothetized that the elevated deep body temperature as a result of a water-saving strategy, was true only for the trunk and not for the brain.

Temperatures and Sweating Rates During Exercise

Evaluation of the monitored variables during the hyperthermia of muscular exercise support the above hypothesis. In dehydrated subjects, the difference between T_{es} and T_{tymp} was enlarged while it disappeared in normally hydrated subjects. Comparably, the difference between sweating rates in dehydration and in control conditions remained high on the back and almost absent on the forehead (Fig. 1 and tables I and II).

Mild hyperthermia of muscular exercise confirmed that the forehead and presumably the face, is a privileged area where sweating is maintained during dehydration while it is inhibited on the rest of the skin surface in order to save water. Thus selective brain cooling can be maintained at least during mild dehydration such as that produced here.

216   *Michel Cabanac*

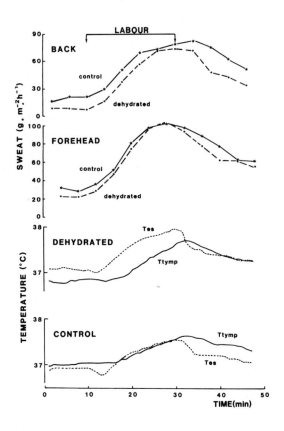

*Fig. 1. Evolution of local sweat secretion and deep body
temperature before, during and after 20 min of exercise on a
cycle ergometer. Each curve is the mean of six sessions (3
subjects taken twice each).* BACK: *sweat rate was recorded
on the back over the lower half of the scapula;* control,
the subjects were normally hydrated; dehydrated, *the subjects
had refrained from ingesting any liquid for at least 24 h.*
FOREHEAD: *sweat rate was recorded on the forehead.*
DEHYDRATED: *oesophageal (T_{es}) and tympanic (T_{tymp}) in the
subjects, dehydrated as defined above.* CONTROL: T_{es} *and
T_{tymp} in the subject normally hydrated (from Cabanac, Caputa
& Massonnet, unpublished).*

It results from the above considerations that the upward
resetting of temperature regulation during dehydration
(Strydom & Holdsnorth, 1968; Nielsen, 1974; Nadel, Fortney

& Wenger, 1980) may be the result of a technical artefact. When plotted against T_{re} or T_{es}, i.e. trunk temperature warm defense response seems to be reset upward. Such resetting was not present in our subjects when sweating rate was plotted against T_{tymp}. Fig. 2 exemplifies this phenomenon; when plotted against itself, T_{tymp} is not much different from the dashed line of no change. If any difference during dehydration, T_{tymp} is lower at rest. On the other hand, T_{es} was about 0.3°C higher during dehydration than in the control condition and thus appeared to be reset upward.

The defense of brain temperature seems to be of paramount importance. After a careful review of the literature, Caputa (1980) estimates that the highest temperature tolerated by the brain is probably 40.5°C. Yet rectal temperature has been recorded to be as high as 41.9°C at the end of a marathon race (Maron, Wagner & Horvath, 1977). It is very likely that selective brain cooling takes place during a race, to protect the brain from these high temperatures. From the above data, one may consider that sweat secretion is maintained on the face even after the intense dehydration of a marathon race. Heat loss from the face must therefore be intense due to the high convection of running. However, when the athletes pass the finish line and rest, one may question how brain temperature evolves if high face convection ends and carotid blood temperature is high. Hirata, Nagasaka & Sugano (1978) have recorded a 0.1°C increase of T_{tymp} at the outset of a short period of mild exercise on an ergometric cycle, i.e. without the air movement of a real exercise. It is therefore of interest to explore T_{tymp} after a prolonged strenuous race or cycling.

INTRACRANIAL TEMPERATURE AFTER AN EXHAUSTING EXERCISE

Nine male well trained subjects in jogging outfit ran on a treadmill at increasing speed and slope until exhaustion. The exercise periods varied with subjects from 33 to 57 min. While each subject was running, a fan was blowing a 14 km h^{-1} wind on his face. The variables T_{es} and T_{tymp} were recorded. At the end of exercise, mean (±S.E.) T_{es} was 39.77±0.07°C and mean T_{tymp} 38.34±0.07°C. Face fanning was therefore capable of maintaining T_{tymp} 1.4°C lower during exercise, a phenomenon already recognized (Cabanac & Caputa, 1979b). Each subject served twice for exactly the same protocol. The difference between the two sessions lay in the 15 minutes of

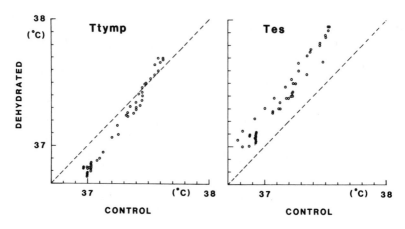

Fig. 2. LEFT: *tympanic temperature in control condition (abscissa) compared to itself during dehydration (ordinate). If dehydration had any influence, it was a decrease of T_{tymp} at rest.* RIGHT: *oesophageal temperature in control condition (abscissa) compared to itself during dehydration (ordinate). Dehydration resulted in a clearcut increase of T_{es} by 0.03-0.4°C (from Cabanac, Caputa & Massonnet, unpublished).*

recovery after the exercise ended. In one session, the fanning ended with the exercise while in the other one, the fan was kept in operation.

Fig. 3 shows the evolution of T_{tymp} during the first 15 min of recovery with and without face fanning. Since T_{tymp} was identical at the end of exercise of both sessions, for each subject, the difference between actual temperature and the last temperature measured during exercise was computed. The curves show the course of the mean of these differences for 8 subjects. It can be seen that when face fanning was interrupted at the end of exercise, mean T_{tymp} rose constantly until 4.5 min after the end of exercise. It reached 38.74±0.14°C. With face fanning, T_{tymp} rose only slightly and plateaued between 1.5 and 4.5 min at 38.44±0.08°C, thereafter decreasing to reach 37.68±0.12 at the end of the 15 min recovery period.

Fanning therefore considerably reduced the initial elevation of T_{tymp} and decreased it much faster. T_{tymp} increased by nearly 0.5°C when face fanning was stopped. This is consistant with the results of Hirata *et al.* (1978). The 0.1°C increase they observed in their subjects was amplified in those of Germain *et al.* (1983) by a more intense physical work

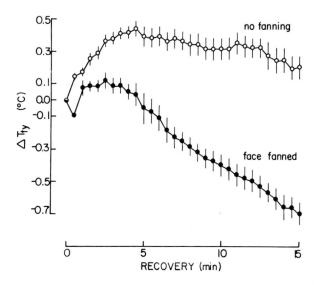

Fig. 3. Tympanic temperature changes (ordinate) follow-ing cessation of exercise at time zero, with (●) and without (○) face fanning. For each subject, the difference between actual temperature and the last temperature recorded during exercise was computed. The curves show the course of the mean (±SE) of these differences for eight subjects (from Germain, Jobin & Cabanac, 1983).

corresponding closely to competitive running. In addition, the fanning provided air convection comparable to that of running or uphill bicycling.

A POSSIBLE THERAPY FOR HEAT STROKE

The high temperature reported in patients suffering from heat stroke may not necessarily reflect brain temperature. It is not easy to measure intracranial temperature especially with subconscious patients. There is good reason to think that actual brain temperature is lower than trunk core temper-ature (the most common measurement), where as high as 46.5°C has been measured in heat stroke (Khogali *et al.*, 1983). Selective brain cooling is an important component of heat defence in humans as well as in other homeotherms. In humans, it can be improved by face fanning.

TABLE I. Sweating: Dehydrated vs Control

	Back		Forehead	
	Rest	*Exercise*	*Rest*	*Exercise*
Paired **t** *test*	2.779	3.386	0.050	0.758
P <	0.025	0.005	N.S.	N.S.

Face fanning is a spontaneous behavioural self defence against hyperthermia. There is a sound physiological basis for this empirical response. Sweating on the forehead is preserved during dehydration and fanning improves evaporative heat loss. When intensive cooling therapy is not available, it may be recommended to simply fan the face of heat stroke subjects and moisten their face. Such very simple therapy may be capable of preventing heat stroke in overheated subjects. Brain temperature can be cooled by at least 1.4°C. Such a difference may be sufficient to avoid dramatic nervous accidents, it would keep the brain of a patient near 38.5°C when his rectal temperature reaches 40°C.

TABLE II. Temperature: T_{es} *vs* T_{tymp}

	Control		Dehydration	
	Rest	*Exercise*	*Rest*	*Exercise*
Paired **t** *test*	0.433	1.551	3.607	18.136
P <	N.S.	N.S.	0.005	0.001

REFERENCES

Adolph, E.F. *et al.* (1947). Physiology of Man in the Desert. Interscience, New York.
Baker, M.A. (1982). Brain cooling in endotherms in heat and exercise. *Ann. Rev. Physiol. 44*, 85-96.
Cabanac, M. & Caputa, M. (1979a). Natural selective cooling of the human brain: Evidence of its occurrence and magnitude. *J. Physiol. (Lond.) 286*, 255-264.

Cabanac, M. & Caputa, M. (1979*b*). Open loop increase in trunk temperature produced by face cooling in working humans. *J. Physiol. (Lond.) 289*, 163–174.

Caputa, M. (1980). Selective brain cooling: An important component of thermal physiology. *In* "Contributions to Thermal Physiology" (Z. Szelényi & M. Székely, eds), pp. 183–192. Akadémiai Kiadó, Budapest.

Caputa, M., Perrin, M. & Cabanac, M. (1978). Ecoulement sanguin réversible dans la veine ophtalmique: Mécanisme de refroidissement sélectif du cerveau humain. *C.R. Acad. Sci. 287*, 1011–1014.

Germain, M., Jobin, M. & Cabanac, M. (1983). Intracranial temperature during the recovery from an exhausting exercise (submitted).

Harrison, M.H. (1974). Plasma volume changes during acute exposure to a high environmental temperature. *J. Appl. Physiol. 37*, 38–42.

Hirata, K., Nagasaka, T. & Sugano, Y. (1978). Effect of alternating respiratory pathway on respiratory capacity, and tympanic and forehead skin temperature during exercise. *Jap. J. Aeorospace Environ. Med. 15*, 8–13.

Khogali, M., El Sayed, H., Amar, M., El Sayad, S., Al Habashi, S. & Mutwali, A. (1983). Management and therapy regimen during cooling and in the recovery room at different heat stroke treatment centres. *In* "Heat Stroke and Temperature Regulation" (M. Khogali & J.R.S. Hales, eds), pp. 149–156. Academic Press, Sydney.

Maron, M.B., Wagner, J.A. & Horvath, S.M. (1977). Thermoregulatory responses during competitive marathon running. *J. Appl. Physiol. 42*, 909–914.

Nadel, E.R., Fortney, S.M. & Wenger, C.B. (1980). Effect of hydration state on circulatory and thermal regulations. *J. Appl. Physiol. 49*, 715–720.

Nielsen, B. (1974). Effect of changes in plasma volume and osmolarity on thermoregulation during exercise. *Acta Physiol. Scand. 90*, 725–730.

Senay, L.C. Jr. (1979). Temperature regulation and hypohydration: A singular view. *J. Appl. Physiol. 47*, 1–7.

Strydom, N.B. & Holdsworth, L.D. (1968). The effect of different levels of water deficit on physiological responses during heat stress. *Int. Z. angew. Physiol. einschl. Arbeitphysiol. 26*, 95–102.

20
Circulatory Consequences of Hyperthermia:
An Animal Model for Studies of Heat Stroke

J. R. S. Hales

C.S.I.R.O.
Ian Clunies Ross Animal Research Laboratory
Prospect (Sydney)
Australia

Thermoregulatory responses to heat place quite specific demands on the cirulatory system, which are met by increasing and/or redistributing cardiac output. This consists principally of increased perfusion of skin, while that of the splanchnic and renal beds is decreased. The skin vasodilatation is greatly reduced with heat stroke, reflecting dominance of cardiovascular requirements over thermoregulation, possibly via low-pressure baroreceptors. Exercise, physical fitness, acclimation, state of hydration, food intake and disease, all greatly influence ability to combat heat stress.

The precise cause of heat stroke is unknown, but it seems likely that some manner of circulatory insufficiency is a factor. Therefore the most suitable animal model for studies of heat stroke might be assessed as one exhibiting a circulatory response to severe hyperthermia most like that of humans. Of the species studied (dog, sheep, baboon), the baboon is an obvious choice. However, the sheep exhibits many similar circulatory responses, and since it is not anthropomorphized like the baboon and is more amenable to experimentation, it should provide a useful animal model.

"Heat stroke" is a complex syndrome where an extremely high body core temperature (>40°C) is usually associated with coma and hot dry skin. Although shock is a critical end-point (Malamud *et al.*, 1946) and the precise cause of heat stroke is unknown, it seems that some manner of circulatory

HEAT STROKE AND
TEMPERATURE REGULATION
ISBN 0 12 406180 X

insufficiency is involved;   circulatory failure precedes
death from heat stroke (Sprung, 1979).   Large variations in
heat flow between deep and superficial tissues are effected
by the cardiovascular system - skin blood flow must increase
if man and other animals are to perform exercise or to survive
as ambient temperature exceeds about 20-25°C.   Based on
measurements under less severe conditions, it is thought that
intense splanchnic vasoconstriction to supply blood volume and
flow to the greatly dilated skin of the heat stressed person,
is followed in the heat stroke patient by cutaneous vasoconstriction in an attempt to avoid drastically reduced central
vascular volume and pressure and consequent shock (see
Shibolet *et al.*, 1976).   As illustrated in Fig. 1, with
progressive hyperthermia in normal subjects, the 6-fold increase in skin blood flow was apparently accomplished by not
only a marked increase in cardiac output, but also by a redistribution of blood flow away from the splanchnic and renal
regions and possibly a little from muscle, i.e., blood was
diverted from vascular beds which assume less importance under
these conditions.

The relatively recent introduction of radioactive micro-
sphere techniques has made it possible to obtain entirely
quantitative information on capillary blood flow rate in most
major tissues of animals.   Figure 2 illustrates the redistribution of cardiac output in conscious sheep at rest and with
progressive hyperthermia in a hot, humid environment.   Blood
flow rate in most tissues was measured (Hales, 1973) but for
simplicity, values for many tissues have been grouped and
expressed as a percentage of cardiac output.   With "moderate
hyperthermia" (rectal temperature raised approximately 1°C)
the total proportion of cardiac output passing through
arteriovenous anastomoses (AVAs) is increased as is capillary
flow in nasobuccal tissues and skin of extremities, presumably all to promote heat loss;   increased blood flow in
respiratory muscles is presumably to meet the energy requirements of panting.   Accompanying this, blood flow is reduced
in what might be regarded under these conditions as 'non-
vital tissues' (abdominal organs and non-respiratory muscles)
and is maintained in the 'vital tissues' (brain, spinal cord
and heart).   It appears that a reservoir of blood for redistribution can also be provided by the fat (Bell *et al.*,
1983).   That is, there is a nicely integrated redistribution
to meet this stress situation.

With the progression to "severe hyperthermia", when
rectal temperature is about 42°C (raised approximately 2.5°C)
but there is not yet any sign of heat stroke in the sheep,
several circulatory responses continue in the same direction

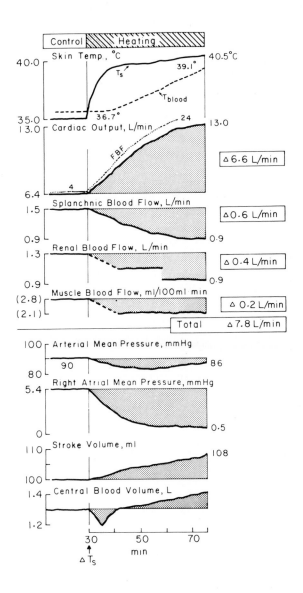

Fig. 1. Cardiovascular adjustments in men heated with a water-perfused suit. T_s = skin temperature, T_{blood} = blood temperature, FBF = forearm blood flow. Compiled by Rowell (1974), courtesy of the American Physiological Society.

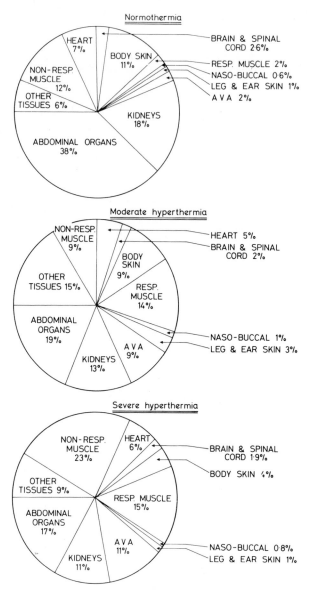

Fig. 2. Microsphere-measured distribution of cardiac output in conscious sheep in a thermoneutral environment and during exposure to a hot humid environment when rectal temperature is raised approximately 1°C ("Moderate") and 2.5°C ("Severe Hyperthermia"). Note that cardiac output, represented by circle area, did not change. From Hales (1976) courtesy of Swets & Zeitlinger.

as initially but others reverse direction. Thus, in heat-
loss tissues the flow decreased towards thermoneutral levels.
For body skin, where earlier there had been no significant
change, flow now decreased (see below). Non-respiratory
muscle, which had decreased, now increased markedly in flow
to almost twice thermoneutral levels. Several responses are
not illustrated, including a progressive decrease in thyroid
blood flow and increase in adrenal blood flow (Hales, 1973,
1974).

In Fig. 3, sheep extremity skin blood flow during severe
hyperthermia is seen to decrease well below levels associated
with efficient control of body temperature during mild heat
stress, and in addition, torso skin flow fell to only 37% of
thermoneutral levels. A "die-away" of flow in human limbs
heated at moderately high levels (38-42.5°C water) has been
described (Thauer, 1965), and for instance, during advanced
stages of exhaustive exercise (Barger *et al.*, 1949) a significant role in the failure to control body temperature is
readily envisaged. It is unlikely that high local skin temperatures are responsible, as although local heating of the
leg to comparable temperatures with the animal in a thermoneutral environment markedly reduces AVA flow (if it is high

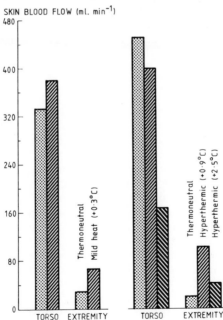

*Fig. 3. Effects of mild and severe heat stress on total
capillary blood flow in skin of conscious sheep. Compiled
from Hales (1973).*

due to some other stimulus) there is actually a massive
enhancement of capillary flow (Hales, 1980;   Hales *et al.*,
1978*a*).   A marked drop in tail temperature of the severely-
stressed, running rat (Hubbard *et al.*, 1976), may also be
evidence of AVA constriction (Hales *et al.*, 1978*b*).   It could
be a consequence of the hypocapnia and alkalosis prevailing
in our panting sheep, or of increased blood levels of
catecholamines (see Hales, 1974;   Robertshaw, 1983).

However, skin vascular activity is clearly an important
part of the shared effector loop of two regulatory systems
(eg. see Nadel, 1983*a,b*;Rowell, 1983), viz., that for blood
pressure as well as that for body temperature.   In this
regard, although Johnson *et al.* (1973) have shown that skin
vasculature of man remains very responsive to arterial baroreceptor input, it has not been possible to make any correlation between the decreased skin blood flow and the small
changes in arterial pressure which occur in heat stressed
animals or man.   However, it is conceivable that the operating
level or sensitivity of arterial baroreceptors could be
modified by thermoregulatory factors, and for instance, even
reduced arterial pulse pressure (without any change in mean)
appears to have a causal relationship with splanchnic and
renal vasoconstriction under some experimental conditions
(eg., lower body negative pressure) (see Rowell, 1983).

It seems more likely that low-pressure baroreceptors,
presumably in the cardiopulmonary region, are involved.
Figure 4 illustrates that lower body negative pressure (used
to mimic haemorrhage in human subjects) does not alter aortic
mean pressure but reduces right atrial pressure.   When this
occurs, very marked splanchnic and even greater forearm
(presumably skin) vasoconstrictions are evoked.   In heat
stressed man, right atrial pressure is reduced (Fig. 1).   In
the anaesthetized, splenectomized dog, the decreased cardiac
output during severe hyperthermia has been attributed to the
fall in central venous pressure (Miki *et al.*, 1983).   The low-
pressure baroreceptors, operating via vagal afferents, can
also lead to renal vasoconstriction, whereas resistance
vessels of muscle are predominantly influenced by arterial
baroreceptors (Pelletier *et al.*, 1971).   It is perhaps, un-
fortunate that the cutaneous veins are not responsive to
either low-pressure (Pelletier *et al.*, 1971) or arterial
baroreceptors (Brender & Webb-Peploe, 1969), as they constitute such a large blood volume and are vitally concerned with
temperature regulation (see Shepherd & Vanhoutte, 1975).
Rowell (1983) has discussed at length, the way in which during
heat stress at rest, central venous pressure is virtually
dominated by activity of the skin vasculature, particularly

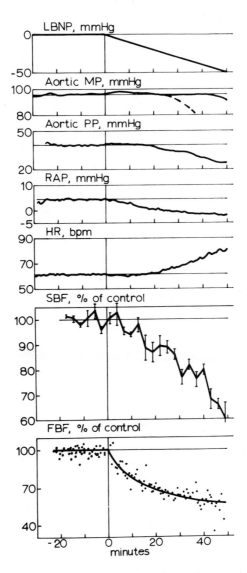

Fig. 4. Average responses of six male subjects to lower body negative pressure (LBNP). MP = mean pressure (broken line is 2 subjects and solid line is the other 4), PP = pulse pressure, RAP = right arterial pressure, HR = heart rate, SBF = splanchnic blood flow, average and S.E., FBF = forearm blood flow, with line fitted by eye. From Johnson et al. (1974), courtesy of the American Heart Association.

the veins - decreased central venous pressure is expected when skin is dilated, ie., when there is increased blood flow to a highly compliant vascular circuit with a long time constant for venous return.   In conformity with this, the importance of differing behaviour of the volumes of various vascular compartments has been illustrated in severely heat stressed dogs: with an unchanged total blood volume, when systemic volume increased, central volume was compelled to decrease (Miki *et al.*, 1983).

There is much evidence for substantial pooling of blood in the periphery, particularly in cutaneous veins (see Rowell, 1983), and recent work has indicated that this normally reduces the body's ability to deal with combined exercise and heat stress.   Thus, when hydrostatic shifts of peripheral venous volumes were prevented (by water immersion), a higher cardiac output and peripheral blood flow could be maintained (Nielsen, 1984).

INFLUENCE OF EXERCISE

Exercise reduces heat tolerance not only because of the additional endogenous heat load, but because of the competition between muscle and skin for circulating blood volume; Rowell (1977*a*) and Nadel (1983*a*) have discussed this.   For example, in exercising man, a maximum oxygen consumption of 3.7 l/min can be attained and is associated with a cardiac output of 22 l/min.   There is a redistribution of blood from splanchnic and renal vascular beds (as with passive heat stress), and from skin, all to meet the metabolically-elicited muscle dilatation.   The peripheral displacement of blood volume is minimized so that central blood volume and arterial pressure are changed little.   When exercise is performed in a hot environment, however, the maximum oxygen consumption may be only 3 l/min and cardiac output 18 l/min, due to reduced stroke volume.   The high skin blood flow initially required to regulate body temperature is progressively reduced, and muscle blood flow supposedly does not reach levels as high as in a thermoneutral environment.

Many of these parameters are estimates and only recently have animal experiments directly demonstrated this competition (Bell *et al.*, 1983).   Note in Fig. 5, that skin does not have such an increase in blood flow with heat exposure during exercise as with heat exposure at rest.   Likewise, exercising skeletal muscle does not receive such an increase in blood flow in a hot as in a thermoneutral environment, ie.,

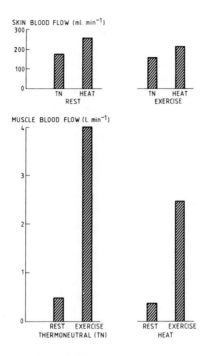

Fig. 5. Effects of exercise and mild heat stress alone and in combination, on total capillary blood flow in muscle (exercising or resting) and skin (torso and limb) of sheep. From Hales et al. (1984), courtesy of Pergamon Press.

exercising muscle may also be a target of sympathetic constrictor activity during exercise in a hot environment. In fact, so must the skin, since the level of either stressor was in itself unchanged when combined with the other stressor, and therefore the dilator drives should not have been reduced. Direct sympathetic nervous stimulation can cause vasoconstriction in contracting muscle (Donald *et al.*, 1970). The inadequacy of the circulating blood volume to supply both the muscles and skin, is not due entirely to failure of a given volume to fill an expanded vascular space. In men, the cutaneous dilatation is also accompanied by loss of both fluid and protein from the vascular volume (Senay, 1972).

INFLUENCE OF STATE OF HYDRATION

Dehydration does not necessarily follow only restricted fluid intake, but develops during heat stress, particularly if combined with exercise, because voluntary fluid intake will not match sweat loss. Effects of such factors on thermal state and cardiovascular function have recently been reviewed (Nadel, 1980, 1983b; Nielsen, 1984), and are considered in this Workshop by Robertshaw (1983) and Appenzeller (1983). Suffice it to say here that with dehydration or hyperosmolality, skin blood flow during exercise does not increase until body core temperature reaches a much higher level than in normal subjects, and the maximum level reached is also reduced (see Robertshaw's Fig. 1). Therefore, a relative hyperthermia is developed. Physical training (Appenzeller, 1983) and heat acclimation (Robertshaw, 1983) should bestow advantages with respect to heat tolerance, not only because of "training effects" on sweating, but because of a greater blood volume to serve cardiovascular requirements.

Alterations in blood volume usually initially have little influence on cardiac output, because reduced cardiac filling and stroke volume are compensated by increased heart rate and peripheral vascular constriction (presumably via the low- and high-pressure baroreceptors) (Nielsen, 1984). With such situations in man, the need for cardiovascular stability takes precedence over the regulation of body temperature. In this regard, the choice of an animal model must be made carefully as there is considerable inter-species variation in the extent of reduction in plasma volume during heat stress and/or dehydration. For example, in the laboratory rat (Horowitz & Borut, 1976) and cat (Schultze et al., 1972) control seems to be as poor as in man (Senay, 1970).

Finally, it is noteworthy that the hyperosmolality and hypernatraemia,which can result from hypo-osmotic sweat loss, may specifically lead to reduced sweat output and consequently aggravation of the hyperthermia.

CONTROL OF REGIONAL BLOOD FLOWS

Several aspects of control, particularly the more "purely cardiovascular", were considered earlier, and no specific attention will be paid to the changes in blood flow to meet local metabolic requirements. However, several studies (Riedel et al., 1974;   Simon & Riedel, 1975) have suggested that the principal redistribution of blood flows during thermal stress is due to differentiation of efferent sympathetic nervous activity in different regions. An input to the "thermoregulatory system", e.g., from warmth receptors, seemingly passes to neural networks at bulbar and suprabulbar levels of the brain stem, which result in widely differing sympathetic outflows, e.g., decrease to skin while there is an increase to the splanchnic bed.

Therefore, an important role for variations in sympathetic constrictor tone is implicated, and in conformity with this, there are reports of major alpha receptor involvement in intestinal vasoconstriction in the hyperthermic baboon (Proppe, 1980) and in the thermally-induced changes in skin blood flow in sheep (Hales *et al.*, 1982). However, this must not be taken as a simple, 'blanket' explanation for the redistribution of cardiac output. At least five considerations are at variance with this concept: (1) It is well established that the major component of skin vasodilatation in response to heat stress in man is due to an active, neurogenic mechanism, ie., increased nerve activity rather than withdrawal of constrictor tone (see Rowell, 1977*b*). (2) Although renal vasoconstriction is associated with increased renal nerve activity in response to heat stimuli (Ninomiya & Fujita, 1976), Eisman & Rowell (1977) prevented most of the renal vasoconstriction in hyperthermic baboons either by blocking release of renin or inhibiting the action of angiotensin II at receptor sites. (3) Many of the blood flow changes still occur after chemical sympathectomy by treatment with six-hydroxydopamine (see Hales, 1983). (4) Evidence is accumulating that responses to catecholamines of arterial smooth muscle in different body regions may vary significantly normally and under the influence of high temperature (Elkhaward *et al.*, 1983). (5) The likely involvement of baroreceptors, as discussed earlier.

A SUITABLE ANIMAL MODEL

Several species, such as dogs (Shapiro *et al.*, 1973) and rats (Hubbard *et al.*, 1977), have been used as models for the study of thermal aspects of heat stroke.

In view of the apparent key role played by the cardiovascular system in determining tolerance to severe heat stress, the most suitable animal model for studies of heat stroke might be assessed as one exhibiting a circulatory

TABLE I.   Effects of hyperthermia on blood flow

	Dog	Sheep	Baboon	Human
Cardiac Output	↑↑	o	o	↑↑
Arterial Pressure	↑	↓	o	o
Extremity Skin	↑↑↑↑	↑↑↑	↑↑↑	↑↑↑↑
Torso Skin	o	↑?	↑	?
Heart	o	o	o	?
CNS	↓	↓	o	?
Respiratory Muscle	↑↑↑	↑↑↑	?	?
Non-Respiratory Muscle	o	↓?	↓	↓?
Kidney	o	↓	↓	↓
Stomach	↓?	↓↓	↓↓ ⎫	
Intestines	↓?	↓↓	↓ ⎬	↓↓
Spleen	↑	↓	↓ ⎭	

? : unknown;   o : no change;   ↑? or ↓? : questionable response.
Arrows indicate responses ranging from small but definite
(↑ or ↓) up to very marked (↑↑↑↑).

response to hyperthermia which is most like that seen, or
thought to occur, in humans.

Table I is a summary of the principal responses of male
Greyhound dogs (Hales & Dampney, 1975), castrated male Merino
sheep (Hales, 1973), male baboons, *Papio anubis* (Hales *et al.*,
1979), and young men (see Rowell, 1974). One should also
bear in mind that within a species, particular breeds or
strains, are likely to show varying responses and even the
environmental conditions under which subjects (human or
animal) have been kept at the time of experimentation could
well be significant. Note that the increased cardiac output
occurring in humans has been seen only in the dog. The stable
arterial pressure in humans is seen only in the baboon.
Extremity skin dilates markedly in all four of the species.
Measurements of the blood flow of torso skin, heart, CNS and
respiratory muscle have not been possible in man. There is a
small decrease in non-respiratory muscle blood flow, both in
man and the sheep. There is a small but definite fall in
kidney blood flow also seen in baboons and sheep but not in
the dog. Finally, the all important splanchnic vasoconstriction seen in man with the splanchnic bed as a whole, is
present in the sub-components which have been measured in
baboons and sheep, but is very doubtful, or even may entail a
response in the opposite direction (spleen), in the dog.

Circulatory responses of the dog are so at variance with
those of man that it should be dismissed as a possible model
(although the splenectomized dog may respond more like man,

Vatner *et al.*, 1974). The baboon seems to be the obvious
model. Except for the surprising lack of any increase in
cardiac output, the circulatory response is very similar to
that in man. Also as in man, there is an active neurogenic
vasodilator mechanism involved in increasing skin blood flow
(Hales *et al.*, 1977). This could be associated with eccrine
sweating, which is present in baboons and humans (Brengelmann
et al., 1981) in contrast to apocrine sweating in most
species, including sheep. Also, the baboon does not pant and
so does not require the increase in respiratory muscle blood
flow necessary in the sheep; nor does it develop the
respiratory alkalosis of panting, which could be involved in
the decreased skin blood flow during severe heat stress and
is probably responsible for the decreased CNS blood flow in
sheep. Finally, the baboon is phylogenetically closer to man.
However, the sheep does exhibit essentially the same pattern

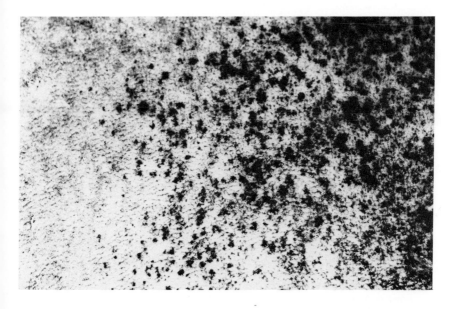

*Fig. 6. Petechiae in midside skin of a sheep suffering
heat stroke. Rectal temperature of 44°C. Fleece had been
removed by treatment 10 days earlier with cyclophosphamide
(orally, 30 mg kg^{-1} fleece-free weight).*

of cardiovascular responses as the baboon, and for example,
if suffering heat stroke it exhibits skin petechiae typical
of man (Fig. 6). I believe the conscious sheep has more
attraction as an experimental animal. We do not see in in
the human-like nature that is apparent in the baboon and it
is certainly much easier to work with in the experimental
laboratory; in particular, the baboon becomes entirely
intractable once body core temperature is raised by more than
approximately 2.5°C. There are also quite significant
factors to be borne in mind regarding availability and cost.
Thus, the sheep could provide a useful animal model.

REFERENCES

Appenzeller, O. (1983). Influences of physical training,
  heat acclimation and diet on temperature regulation in
  man. *In* "Heat Stroke and Temperature Regulation" (M.
  Khogali & J.R.S. Hales, eds), pp.283-292. Academic Press,
  Sydney.
Barger, A.C., Greenwood, W.F., Dipalma, J., Stokes, J. &
  Smith, L. (1949). Venous pressure and cutaneous reactive
  hyperemia in exhausting exercise and certain other
  circulatory stress. *J. Appl. Physiol. 2*, 81-96.
Bell, A.W., Hales, J.R.S., King, R.B. & Fawcett, A.A. (1983).
  Influences of heat stress on exercise-induced changes in
  regional blood flow in sheep. *J. Appl. Physiol.*
  (in press).
Brender, D. & Webb-Peploe, M. (1969). Influence of carotid
  baroreceptors on different components of the vascular
  system. *J. Physiol. 205*, 257-274.
Brengelmann, G.L., Freund, P.R., Rowell, L.B., Olerud, J.E. &
  Kraning, K.K. (1981). Absence of active cutaneous vasodilatation associated with congenital absence of sweat
  glands in humans. *Am. J. Physiol. 240*, H571-575.
Donald, D.E., Rowlands, D.J. & Ferguson, D.A. (1970).
  Similarity of blood flow in the normal and the sympathectomized dog hind limb during graded exercise. *Cir. Res.
  26*, 185-199.
Eisman, M.M. & Rowell, L.B. (1977). Renal vascular response
  to heat stress in baboons - role of renin-angiotensin.
  J. Appl. Physiol. 43, 739-746.

Elkhawad, A., Khogali, M., Thulesius, O., Hamdan, S.M. &
    Hassan, O. (1983). Interaction of hyperthermia in blood
    vessels. *In* "Heat Stroke and Temperature Regulation"
    (M. Khogali & J.R.S. Hales, eds), pp.263-271. Academic
    Press, Sydney.
Hales, J.R.S. (1973). Effects of exposure to hot environ-
    ments on the regional distribution of blood flow and on
    cardiorespiratory function in sheep. *Pflügers Arch.*
    344, 133-148.
Hales, J.R.S. (1974). Physiological responses to heat. *In*
    "MTP International Reviews of Science : Environmental
    Physiology" (D. Robertshaw, ed.), pp. 107-162.
    Butterworths, London.
Hales, J.R.S. (1976). The redistribution of cardiac output
    in animals during heat stress. *In* "Progress in Bio-
    meteorology, Ser. B. Animal Biometeorology" (H.D.
    Johnson, ed.), Vol. 1, pp. 285-294. Swets & Zeitlinger,
    Amsterdam.
Hales, J.R.S. (1980). Paradoxical effects of temperature on
    skin arteriovenous anastomoses. *In* "Contributions to
    Thermal Physiology" (Z. Szelényi & M. Székely, eds),
    pp. 383-385. Akadémiai Kiadó, Budapest.
Hales, J.R.S. (1983). Thermoregulatory requirements for
    circulatory adjustments to promote heat loss in animals:
    A review. *J. Thermal Biol. 8*, 219-224.
Hales, J.R.S., Bell, A.W., Fawcett, A.A. & King, R.B. (1984).
    Redistribution of cardiac output and skin AVA activity
    in sheep during heat stress and exercise. *J. Thermal*
    Biol. (in press).
Hales, J.R.S. & Dampney, R.A.L. (1975). The redistribution
    of cardiac output in the dog during heat stress.
    J. Thermal Biol. 1, 29-34.
Hales, J.R.S., Fawcett, A.A., Bennett, J.W. & Needham, A.D.
    (1978*a*). Thermal control of blood flow through
    capillaries and arteriovenous anastomoses in skin of
    sheep. *Pflügers Arch. 378*, 55-63.
Hales, J.R.S., Foldes, A., Fawcett, A.A. & King, R.B. (1982).
    The role of adrenergic mechanisms in thermoregulatory
    control of blood flow through capillaries and arterio-
    venous anastomoses in the sheep hind limb. *Pflügers*
    Arch. 395, 93-98.

Hales, J.R.S., Iriki, M., Tsuchiya, K. & Kozawa, E. (1978*b*).
Thermally-induced cutaneous sympathetic activity related
to blood flow through capillaries and arteriovenous
anastomoses. *Pflügers Arch. 375*, 17–24.

Hales, J.R.S., Rowell, L.B. & King, R.B. (1979). Regional
distribution of blood flow in awake, heat-stressed
baboons. *Am. J. Physiol. 237*, H705–712.

Hales, J.R.S., Rowell, L.B. & Strandness, D.E. (1977).
Active cutaneous vasodilatation in the hyperthermic
baboon. *Proc. Aust. Physiol. Pharmacol. Soc. 8*, 70.

Horowitz, M. & Borut, A. (1976). Plasma volume regulation in
rodents. *Israel J. Med. Sci. 12*, 864–867.

Hubbard, R.W., Bowers, W.D. & Mager, M. (1976). A study of
physiological, pathological and biochemical changes in
rats with heat- and/or work-induced disorders. *Israel
J. Med. Sci. 12*, 884–886.

Hubbard, R.W., Bowers, W.D., Matthew, W.T., Curtis, F.C.,
Criss, R.E.L., Sheldon, G.M. & Ratteree, J.W. (1977).
Rat model of acute heatstroke mortality. *J. Appl.
Physiol. 42*, 809–816.

Johnson, J.M., Niederberger, M., Rowell, L.B., Eisman, M.M. &
Brengelmann, G.L. (1973). Competition between cutaneous
vasodilator and vasoconstrictor reflexes in man.
J. Appl. Physiol. 35, 798–803.

Johnson, J.M., Rowell, L.B., Niederberger, M. & Eisman, M.M.
(1974). Human splanchnic and forearm vasoconstrictor
responses to reductions of right atrial and aortic
pressures. *Circ. Res. 34*, 515–524.

Malamud, N., Haymaker, W. & Custer, R.P. (1946). Heatstroke:
A clinicopathologic study of 125 fatal cases. *Milit.
Surg. 99*, 397–449.

Miki, K., Morimoto, T., Nose, H., Itoh, T. & Yamada, S.
(1983). Circulatory failure during severe hyperthermia
in dog. *Jap. J. Physiol. 33*, 269–278.

Nadel, E.R. (1980). Circulatory and thermal regulation during exercise. *Fed. Proc. 39*, 1491–1497.

Nadel, E.R. (1983*a*). Factors affecting the regulation of
body temperature during exercise. *J. Thermal Biol.
8*, 165–169.

Nadel, E.R. (1983*b*). Cardiovascular, body fluid and electrolyte balance during exercise and heat: competing demands
with temperature regulation. *In* "Thermal Physiology"
(J.R.S. Hales, ed.), in press. Raven Press, New York.

Nielsen, B. (1984). The effect of dehydration on circulation
and temperature regulation during exercise. *J. Thermal
Biol.* (in press).

Ninomiya, I. & Fujita, S. (1976). Reflex effects of thermal stimulation on sympathetic nerve activity to skin and kidney. *Am. J. Physiol. 230*, 271-278.

Pelletier, C.L., Edis, A.J. & Shepherd, J.T. (1971). Circulatory reflex from vagal afferents in response to hemorrhage in the dog. *Circ. Res. 29*, 626-634.

Proppe, D.W. (1980). α-Adrenergic control of intestinal circulation in heat-stressed baboons. *J. Appl. Physiol. 48*, 759-764.

Riedel, W., Iriki, M. & Simon, E. (1974). Functional variability of regional qualitative differentiation of sympathetic outflow. *In* "Central Rhythmic and Regulation" (W. Umbach & H.P. Koepchen, eds), pp. 228-234. Hippokrates, Stuttgart.

Robertshaw, D. (1983). Contributing factors to heat stroke. *In* "Heat Stroke and Temperature Regulation" (M. Khogali & J.R.S. Hales, eds), pp. 13-29. Academic Press, Sydney.

Rowell, L.B. (1974). Human cardiovascular adjustments to exercise and thermal stress. *Physiol. Rev. 54*, 75-159.

Rowell, L.B. (1977*a*). Competition between skin and muscle for blood flow during exercise. *In* "Problems with Temperature Regulation during Exercise" (E.R. Nadel, ed.), pp. 49-76. Academic Press, New York.

Rowell, L.B. (1977*b*). Reflex control of the cutaneous vasculature. *J. Invest. Dermatol. 69*, 154-166.

Rowell, L.B. (1983). Cardiovascular adjustments to thermal stress. *In* "Handbook of Physiology - The Cardiovascular System" (J.P. Shepard & F.M. Abboud, eds), in press. Am. Physiol. Soc., Bethesda.

Schultze, G., Kirsch, K. & Röcker, L. (1972). Distribution and circulation of extracellular fluid and protein during different states of hydration in the cat. *Pflügers Arch. 337*, 351-366.

Senay, L.C. (1970). Movement of water, protein and crystalloids between vascular and extravascular compartments in heat-exposed men during dehydration and following limited relief of dehydration. *J. Physiol. 210*, 617-635.

Senay, L.C. (1972). Changes in plasma volume and protein content during exposures of working men to various temperatures before and after acclimatization to heat: separation of the roles of cutaneous and skeletal muscle circulation. *J. Physiol. 224*, 61-81.

Shapiro, Y., Rosenthal, T. & Sohar, E. (1973). Experimental heatstroke. A model in dogs. *Arch. Intern. Med. 131*, 688-692.

Shepherd, J.J. & Vanhoutte, P.M. (1975). "Veins and Their Control." Saunders, London.

Shibolet, S., Lancaster, M.C. & Danon, Y. (1976).   Heat
    Stroke: A review.   *Aviat. Space Environ. Med. 47*, 280–301.

Simon, E. & Riedel, W. (1975).   Diversity of regional
    sympathetic outflow in integrative cardiovascular control:
    patterns and mechanisms.   *Brain Res. 87*, 323–333.

Sprung, C.L. (1979).   Haemodynamic alterations of heat stroke
    in the elderly.   *Chest 75*, 362–366.

Thauer, R. (1965).   Circulatory adjustments to climatic
    requirements.   *In* "Handbook of Physiology – Circulation"
    Sec. 2 (W.F. Hamilton & P. Dow, eds), Vol. III, pp. 1921–
    1966.   Am. Physiol. Soc., Washington, D.C.

Vatner, S.F., Higgins, C.B., Millard, R.W. & Franklin, D.
    (1974).   Role of the spleen in the peripheral vascular
    response to severe exercise in untethered dogs.
    Cardiovasc. Res. 8, 276–282.

21
Some Aspects of Cutaneous Blood Flow and Acid–Base Balance during Hyperthermia

Claus Jessen* and Gero Feistkorn

Physiologisches Institut
Universität Giessen
Germany

The afferent input to the temperature regulating system, which drives the heat dissipating mechanisms during hyperthermia, is generated by thermosensitive structures distributed throughout the body. Using a newly developed animal preparation it could be shown that cutaneous vasodilatation is effected to the same extent by local hyperthermia of the head or hyperthermia confined to the trunk.

Cutaneous heat loss at the extremities is a function of blood flow through capillaries and arteriovenous anastomoses (AVAs). Recent experiments in sheep indicate that high levels of AVA flow, induced by high levels of body core temperature, can be further enhanced by clamping skin temperature moderately below core temperature, thereby greatly facilitating heat loss.

If hyperthermia in a panting animal occurs during exercise a primary respiratory alkalosis is partially compensated by a metabolic acidosis, which is caused by an increase in blood lactate concentration. At constant work rate, the lactate concentration is a function of body core temperature. It may contribute to exhaustion, which is linked to the accumulation of lactic acid in the exercising muscle, and occurs at constant work rate earlier with higher levels of body temperature.

*Supported by DFG Je 57/8-5

HEAT STROKE AND
TEMPERATURE REGULATION
ISBN 0 12 406180 X

CUTANEOUS BLOOD FLOW

In this paper hyperthermia is referred to as the
condition of an intact temperature regulating animal when
core temperature is more than one standard deviation above
the mean of the species in resting conditions in a thermoneutral environment and all mechanisms of heat dissipation
are activated. This includes cutaneous vasodilatation which
is instrumental in permitting a sufficient flow of heat from
body core to surface. Concerning the efferent part of the
system mediating cutaneous vasodilatation, there is general
agreement that blood flow through the skin of the
extremities is controlled exclusively via noradrenergic
sympathetic fibres: a decrease of the sympathetic tone
causes vasodilatation. In man, skin blood flow in the trunk
and in the proximal limbs can be increased by heat-induced
active reflex vasodilatation. These cutaneous regions are
the same in which pronounced thermal sweating occurs.
Sweating and active vasodilatation are closely associated,
although the mechanisms which link sudomotor and vasodilator
activities in the skin are far from being understood
(Brengelmann *et al.*, 1981). Furthermore, local effects of
skin temperature on skin blood flow have to be taken into
account: the skin of isolated and denervated limb areas
continues to respond to thermal stimuli by vasomotor adjustments (Pappenheimer *et al.*, 1948; Perkins *et al.*, 1948).
Concerning the afferent control of skin blood flow, a
number of linear and nonlinear models with mean skin temperature and core temperature as inputs have been described
(Wyss *et al.*, 1975). Depending on which central temperature
is taken to represent body core temperature the relative
influence of core temperature in control of skin blood flow
can exceed that of mean skin temperature by a factor of 20
(Wyss *et al.*, 1974). As to the location of thermosensitive
elements within the body core, the anterior hypothalamic
centre or "temperature eye" has long been considered the
unique site in which all thermoreceptive functions of the
inner body are concentrated (Benzinger, 1969). However, in
1964 Simon *et al.* showed that local cooling of the spinal
cord in the anaesthetized dog induces shivering and a rise
in heat production (Simon *et al.*, 1964). In the conscious
dog, local warming of the spinal cord was shown to elicit
cutaneous vasodilatation (Jessen *et al.*, 1967). The
discovery of spinal temperature sensitivity has stimulated
investigations for other extrahypothalamic temperature
sensors, and at present it appears to be well established

that, beside the hypothalamus and the spinal cord, the medulla oblongata (Chai & Lin, 1972; Lipton, 1973) and the midbrain (Schmieg *et al.*, 1980) contain temperature sensing elements the signals of which feed into the controller of body temperature. However, temperature sensors are not restricted to the central nervous system: they have also been found in the abdomen (Rawson & Quick, 1972) close to the splanchnic root (Riedel *et al.*, 1973). More recently, evidence has been provided for thermal afferents from skeletal muscle (Jessen *et al.*, 1983).

Some of these studies have encountered criticism on grounds of the methods which were employed to trace the thermosensitive structures: all experiments involved local thermodes and necessitated strong thermal stimuli, creating highly unnatural thermal gradients within the body core. Despite all the evidence cited above, it was recently concluded that in working humans, thermoregulatory processes relate only to brain temperature and not to body temperature generally (Cabanac & Caputa, 1979). Therefore a new method has been developed recently to assess thermal sensitivity of head and trunk independently of each other in the conscious goat (Jessen *et al.*, 1982). In a sterile operation, both vertebral arteries were tied and extra-corporeal loops made from silicone tubes inserted in the common carotid arteries. The tubes could be connected to heat exchangers to provide control of head temperature. Independent manipulation of trunk temperature was achieved by means of chronically implanted heat exchangers in the large veins (Jessen, 1981). The largest temperature difference which could be sustained between head and trunk amounted to 6.6°C. Apart from heat production and respiratory evaporative heat loss, skin temperatures at an ear and a foreleg were continuously recorded as an index of skin blood flow.

For both skin sites, Fig. 1 shows those combinations of head temperature and trunk temperature, at which an abrupt alteration of skin temperature was indicative of a change from vasoconstriction to vasodilatation, and vice versa. Just above the lines, ear skin temperature was on the average, 6.0±2.0°C higher than below the line. At the fore-leg, the increment amounted to 2.9±0.8°C. The shape of the lines joining the points is indicative of skin blood flow being controlled by thermal inputs generated in the head and in the trunk: if trunk temperature exceeded 40.5°C, vaso-dilatation occurred regardless of head temperature being as low as 36.5°C. Head temperature above 40.0°C invariably induced vasodilatation even if trunk temperature

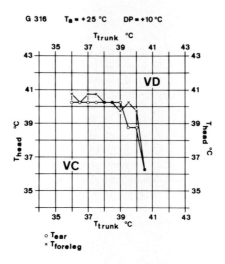

Fig. 1. Combinations of trunk temperature and head temperature, when steep changes of skin temperature at an ear and a foreleg indicated cutaneous vasoconstriction (VC) or vasodilatation (VD). Mean values of 2990 measurements in 16 experiments in a conscious goat.

was as low as 36.0°C. Similar relationships were found for heat production and respiratory evaporative heat loss. Thus, available evidence strongly suggests discarding the hypothesis of a unique thermosensitive centre of the hypothalamus, and replacing it by the concept of a general distribution of thermosensitive elements throughout the body, all feeding signals into a common controlling system (Mitchell, 1972).

In many mammalian species the skin of the extremities is richly supplied with arteriovenous anastomoses (AVAs) that connect the arterial and venous sides of the vascular bed and allow arterial blood to enter venous circulation without passing through the capillary bed. Due to their size and the location in the superficial layers of the skin, the AVAs can handle the major part of the cutaneous blood flow and there- fore mainly influence the magnitude of dry heat loss. In a recent study on conscious sheep Hales *et al*. (1978) were able to determine AVA flow and capillary flow in the skin of a hindleg independently of each other, while the animal was

subjected to various heating treatments. Local heating of
the hypothalamus, the spinal cord or the forelegs increased
AVA flow and left capillary flow unchanged, while local
warming of the hindleg increased capillary flow and did not
affect AVA flow. It was concluded that blood flow through
cutaneous AVAs is the target of specific thermoregulatory
reflexes, whereas capillary flow is subject to direct local
effects of temperature. Concerning the efferent neural
control of the AVAs its adrenergic nature is well established, although there is some evidence for an additional
nonadrenergic nerve supply (Molyneux, 1981).

The patency of cutaneous AVAs appears to be further
controlled by the direction of the thermal gradient across
the skin. In a series of pilot experiments high cutaneous
AVA flow induced either by spinal cord warming or by external
heat, decreased to near control levels when local warming of
the hindlegs by water at 44°C was superimposed (Hales,
1980*a*). This was teleologically interpreted as preventing
heat from flowing into the body when the normal gradient is
reversed. However, a recent series of experiments in
conscious sheep suggests a more complex relationship between
internal and local external temperatures in control of AVA
flow (Hales *et al.*, 1983).

In 5 conscious sheep 21 experiments were performed at air
temperatures of 15 and 30°C. The animals were equipped with
chronically implanted intravascular heat exchangers,
permitting control of body core temperature between 38 and
42°C independently of ambient temperature (Jessen, 1981). A
chronically implanted electromagnetic flow probe around the
femoral artery and chronic catheters in the femoral and
pulmonary arteries were used to determine AVA flow in a hindleg (Hales *et al.*, 1978). In 9 experiments body core temperature was clamped at 40 or 41°C, while the skin temperature
at the hindleg was locally varied, by means of a water bath,
between 36 and 44°C. AVA flow attained a minimum at a leg
skin temperature of 40°C, i.e. at or near the zero thermal
gradient across the skin. With rising skin temperature, AVA
flow showed a gradual increase. However, the most
conspicuous effect on AVA flow was seen when leg skin temperature was lowered: AVA flow increased with falling skin
temperature and reached its maximum at a skin temperature of
36°C. This relationship with local skin temperature was
regularly observed when the animal was in a hyperthermic
state and its heat dissipating mechanisms were activated.
However, the higher body temperature was, the more
pronounced was the response. Fig. 2 compares the results of
two experiments in one animal at 40 and 41°C core

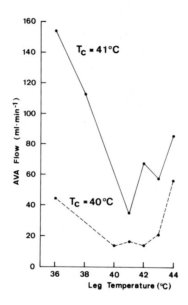

Fig. 2.   Hindleg AVA flow at different levels of hindleg surface temperature, when general body core temperature (T_C) was 40°C or 41°C.   2 experiments in a conscious sheep (Hales et al., 1983)

temperature. In both experiments AVA flow was at its minimum near the zero gradient across the skin and rose with skin temperature rising or falling. However, at any skin temperature the absolute levels of AVA flow were higher at higher core temperature. It may be noted that the highest AVA flow which was observed in this series (155 ml/min at 41°C core temperature and 36°C skin temperature) amounted to more than 90% of the total femoral flow.

The results of these experiments are in agreement with previous findings insofar as it could be confirmed that AVA flow is not entirely defined by the general thermal state of the body but is also under local control of skin temperature. Furthermore, the precise control of body core temperature, which was exerted in these experiments, and the use of discrete steps in skin temperature have resolved that AVA flow during general hyperthermia attains its minimum at a zero thermal gradient across the skin. With higher and even more with lower skin temperature, AVA flow increases.

Similar observations have been made in the dog's tongue (Krönert *et al.*, 1980): AVA flow more than tripled when the temperature of the surface of the tongue was lowered from 40.5 to 27.7°C. Since complete denervation did not abolish this response, a reflex mechanism was ruled out and the authors have suggested that the smooth muscle fibres of the AVAs could be controlled by axon collaterals to skin cold receptors, although there is no histological evidence for that (Molyneux, 1981). Whatever the underlying mechanisms may be, the phenomenon deserves further exploration because of its possible practical consequences. The high AVA flow, which is induced by the combining effects of general hyperthermia and skin temperature moderately below body temperature, greatly facilitates heat loss and provides a powerful heat sink for the body.

ACID-BASE BALANCE

In severe hyperthermia, induced in a panting species by a large external heat load, acid-base balance is greatly disturbed by alveolar ventilation exceeding metabolic requirements. Arterial PCO_2 may decline to less than 10 Torr and pH may exceed 7.7: this is the typical picture of respiratory alkalosis, prevailing during "second-phase panting" (Hales, 1980*b*). However, if hyperthermia is induced by a combination of an external heat load and exercise, the respiratory alkalosis is reduced by a higher production of CO_2. In a recent series of experiments the effects on acid-base balance of hyperthermia during exercise were studied in conscious goats (Feistkorn *et al.*, 1982). The animals were equipped with chronically implanted heat exchangers permitting control of body temperature independently of work rate. In a hot and dry environment, body core temperature was adjusted to 39°C or 42°C and maintained at these levels for nearly two hours. During the last 60 min the animals worked at a rate of 1.22 W/kg (treadmill, 3 km/h, slope 15%). Table 1 shows mean values and standard deviations of arterial PCO_2 (Torr) arterial pH and arterial lactate (mmol/l) at selected instants of 2x5 experiments in 2 animals.

The first row (0_1 min) shows the values obtained at the very beginning of the experiments, i.e. at unclamped body temperature. All values are within the normal range. The second row (-11) contains data taken shortly before exercise was started, i.e. approximately 50 min after body temperature

TABLE 1.   *Arterial PCO_2 (Torr), pH, and lactate concentration (mmol/L) at normal body temperature (0_1), at body temperatures of 39°C or 42°C during rest (-11) and during exercise ($+13$, $+27$, $+55$) on a treadmill (speed 3 km/h, slope 15%) at an air temperature of 35°C and a relative humidity of 33%. Mean values, standard deviations of, and t-tests between, five experiments at each condition in two goats (G 300 and G 310).*

t min		G 300			G 310		
		39.0°C	42.0°C		39.0°C	42.0°C	
0_1	PCO_2	41.8±1.6	41.3±1.6	n.s.	39.1±0.9	37.8±0.5	n.s.
	pH	7.40±.02	7.39±.02	n.s.	7.36±.01	7.39±.03	n.s.
	LA	0.64±0.1	0.48±0.1	n.s.	0.41±0.1	0.50±0.0	n.s.
-11	PCO_2	41.9±0.9	26.4±2.5	$p<0.01$	40.6±1.1	19.8±1.6	$p<0.01$
	pH	7.38±.02	7.55±.02	$p<0.01$	7.37±.03	7.59±.10	$p<0.05$
	LA	0.60±0.1	1.13±0.5	n.s.	0.47±0.1	2.29±0.7	$p<0.01$
$+13$	PCO_2	38.8±1.1	27.6±5.3	$p<0.01$	36.7±2.2	21.5±1.0	$p<0.01$
	pH	7.38±.01	7.45±.04	$p<0.05$	7.34±.01	7.51±.03	$p<0.01$
	LA	2.23±1.0	5.59±1.5	$p<0.01$	2.35±0.9	7.96±0.4	$p<0.01$
$+27$	PCO_2	37.6±0.8	27.6±5.0	$p<0.01$	36.3±1.3	22.0±0.8	$p<0.01$
	pH	7.40±.01	7.47±.04	$p<0.05$	7.37±.02	7.48±.03	$p<0.01$
	LA	1.83±0.6	4.92±0.8	$p<0.01$	2.00±1.1	7.35±1.0	$p<0.01$
$+55$	PCO_2	37.1±0.9	27.2±3.5	$p<0.01$	37.7±1.4	23.0±0.6	$p<0.01$
	pH	7.41±.02	7.49±.04	$p<0.01$	7.39±.01	7.45±.06	n.s.
	LA	1.88±0.5	4.32±1.0	$p<0.01$	1.05±0.3	6.29±1.1	$p<0.01$

was adjusted to either 39°C or 42°C. Therefore, the 42°C values show the effects of severe hyperthermia during rest. Typically, PCO_2 was decreased and pH was increased. This was accompanied by a slight increase of the lactate concentration, which rose from 0.5 to 1.1 and 2.3 mmol/l in the two goats.

The last four rows contain the data obtained 13, 27, 41, and 55 min after start of exercise. At a normal body temperature of 39°C, neither PCO_2 nor pH showed any significant deviation from resting values. There was a moderate increase of lactate concentration, which reached a maximum at the early stage of exercise and then gradually declined. In the hyperthermic state, arterial PCO_2 was virtually unchanged as compared to rest. The previously alkalotic pH however, was nearly restored to its normal value. This was due to a marked increase in lactate concentration, which attained values as high as 8 mmol/l. Thus, the acid-base state of a hyperthermic panting animal during exercise is characterized by a respiratory alkalosis compensated by a metabolic acidosis. The higher levels of arterial lactate were linked to hyperthermia and did not occur when the same work rate was performed at normothermia.

High arterial lactate concentrations have been observed in a number of species during hyperthermia and graded stages of heat stroke. In oxen exposed to severe heat, excess lactate was found to develop when rectal temperature exceeded 41°C (Hales *et al.*, 1967). Dogs exposed to an ambience of 45°C and 25% RH developed a heat stroke, when rectal temperature reached 44.1°C (Magazanik *et al.*, 1980). On the way to this level, an initial respiratory alkalosis was followed by a severe metabolic acidosis, which occurred above 42°C rectal temperature. At the peak, arterial pH was below 7.2 and base excess was less than -10 mEq/l. Again the acidosis was caused by an increase in lactate concentration, reaching approximately 7 mmol/l.

Also in man, either during experimental hyperthermia or during varying degrees of heat illness or heat stroke, metabolic acidosis caused by an increase in arterial lactate concentration, is a most common observation (Eichler *et al.*, 1969; Ruppert *et al.*, 1964; Shibolet *et al.*, 1976).

An increase in blood lactate concentration is usually taken as indicative of anaerobic metabolism. However, it is usually combined with high arterial PO_2. This led Frankel & Ferrante (1966) to assume that it is the decrease in PCO_2 and a primary respiratory alkalosis, induced by hyperthermic hyperventilation, which precedes and causes blood lactate concentration to rise. Alkalosis has been found to increase the lactate permeation rate from exercising muscles, thereby increasing blood lactate and decreasing intracellular lactate at a virtually unchanged level of lactate production (Hirche *et al.*, 1975). However, respiratory alkalosis and appearance of excess blood lactate are not necessarily parallel to each other (Hales *et al.*, 1967). At least in exercising sheep, even mild hyperthermia has been reported to reduce the increase in blood flow to fore- and hindlimb muscles (Bell *et al.*, 1983). Thus, it may well be that in exercising muscles during hyperthermia, local hypoxia develops with a subsequent increase of anaerobic metabolism. This could be the primary cause of metabolic acidosis during hyperthermia and could contribute to exhaustion, which is linked to the accumulation of lactic acid in the exercising muscle and occurs, at constant work rate, earlier with high levels of body temperature.

REFERENCES

Bell, A.W., Hales, J.R.S., King, R.B. & Fawcett, A.A. (1983).
    Influences of heat stress on exercise-induced changes in
    regional blood flow in sheep. *J. Appl. Physiol.*
    (in press).
Benzinger, T.H. (1969). Heat regulation: homeostasis of
    central temperature in man. *Physiol. Rev. 49*, 671-759.
Brengelmann, G.L., Freund, P.R., Rowell, L.B., Olerud, J.E.
    & Kraning, K.K. (1981). Absence of active cutaneous
    vasodilation associated with congenital absence of sweat
    glands in humans. *Am. J. Physiol. 240*, H 571-575.
Cabanac, M. & Caputa, M. (1979). Open loop increase in trunk
    temperature produced by face cooling in working humans.
    J. Physiol. 289, 163-174.
Chai, C.Y. & Lin, M.T. (1972). Effects of heating and cool-
    ing the spinal cord and medulla oblongata on thermo-
    regulation in monkeys. *J. Physiol. 225*, 297-308.
Claremont, A.D., Nagle, F., Reddan, W.D. & Brooks, G.A.
    (1975). Comparison of metabolic, temperature, heartrate
    and ventilatory responses to exercise at extreme ambient
    temperatures (0°C and 35°C). *Med. Sci. Sports
    7*, 150-154.
Eichler, A.C., McFee, A.S. & Root, H.D. (1969). Heat stroke.
    Am. J. Surg. 118, 855-863.
Feistkorn, G., Nagel, A. & Jessen, C. (1982). Effects of
    body temperature on cardiovascular function in the
    exercising goat. *Pflügers Arch., 394, Suppl.*, R 38.

Frankel, H.M. & Ferrante, F.L. (1966). Effect of arterial pCO$_2$ on appearance of increased lactate during hyperthermia. *Am. J. Physiol. 210*, 1269-1272.

Hales, J.R.S. (1980*a*). Paradoxical effects of temperature on skin arteriovenous anastomoses. *In* "Contributions to Thermal Physiology" (Z. Szelény & M. Székely, eds), pp. 383-385. Akadémiai Kiadó, Budapest.

Hales, J.R.S. (1980*b*). Peripheral effector mechanisms of thermoregulation. Regulation of Panting. *In* "Contributions to Thermal Physiology" (Z. Szelény & M. Székely, eds), pp. 421-425. Akadémiai Kiadó, Budapest.

Hales, J.R.S., Fawcett, A.A., Bennett, J.W. & Needham, A.D. (1978). Thermal control of blood flow through capillaries and arteriovenous anastomoses in skin of sheep. *Pflügers Arch. 378*, 55-63.

Hales, J.R.S., Findlay, J.D. & Mabon, R.M. (1967). Tissue hypoxia in oxen exposed to severe heat. *Resp. Physiol. 3*, 43-46.

Hales, J.R.S., Jessen, C., King R.B. & Fawcett, A.A. (1983). Role of local skin temperature in control of arteriovenous anastomoses (in preparation).

Hirche, H., Hombach, V., Langohr, D., Wacker, W. & Busse, J. (1975). Lactic acid permeation rate in working gastrocnemii of dogs during metabolic alkalosis and acidosis. *Pflügers Arch. 356*, 209-222.

Jessen, C. (1981). Independent clamps of peripheral and central temperatures and their effects on heat production in the goat. *J. Physiol. 311*, 11-22.

Jessen, C., Feistkorn, G. & Nagel, A. (1982). Thermal separation of head and trunk in the conscious goat. *Pflügers Arch. 394, Suppl.*, R 40.

Jessen, C., Feistkorn, G. & Nagel, A. (1983). Temperature sensitivity of skeletal muscle in the conscious goat. *J. Appl. Physiol.* (in press).

Jessen, C., Meurer, K.A. & Simon, E. (1967). Steigerung der Hautdurchblutung durch isolierte Warmung des Ruckenmarks am wachen Hund. *Pflügers Arch. 297*, 35-52.

Krönert, H., Wurster, R.D., Pierau, Fr.-K. & Pleschka, K. (1980). Vasodilatory response of arteriovenous anastomoses to local cold stimuli in the dog's tongue. *Pflügers Arch. 388*, 17-19.

Lipton, J.M. (1973). Thermosensitivity of medulla oblongata in control of body temperature. *Am. J. Physiol. 224*, 890-897.

Magazanik, A., Shapiro, Y. & Shibolet, S. (1980). Dynamic changes in acid base balance during heat stroke in dogs. *Pflügers Arch. 388*, 129-135.

Mitchell, D. (1972). Human surface temperature: its measurement and its significance in thermoregulation. Thesis, University of Witwatersrand, Johannesburg, South Africa.

Molyneux, G.S. (1981). Neural control of cutaneous arteriovenous anastomoses. *In* "Progress in Microcirculation Research" (D. Garlick, ed.), pp. 296-315. Committee in Postgraduate Med. Ed. U.N.S.W., Sydney, Australia.

Pappenheimer, J.R., Eversole jr., S.L. & Soto-Rivera, A. (1948). Vascular responses to temperature in the isolated perfused hindlimb of the cat. *Am. J. Physiol.* *155*, 458.

Perkins, jr., J.F., Li, M.C., Hoffman, F. & Hoffmann, E. (1948). Sudden vasoconstriction in denervated or sympathectomized paws exposed to cold. *Am. J. Physiol.* *155*, 165-178.

Rawson, R.O. & Quick, K.P. (1972). Localization of intra-abdominal thermoreceptors in the ewe. *J. Physiol.* *222*, 665-677.

Riedel, W., Siaplauras, G. & Simon, E. (1973). Intra-abdominal thermosensitivity in the rabbit as compared with spinal thermosensitivity. *Pflügers Arch.* *340*, 59-70.

Ruppert, R.D., Newman, A., Scarpelli, D.G. & Weissler, A.M. (1964). The mechanisms of metabolic acidosis in heat stroke. *Clin. Res.* *12*, 356.

Schmieg, G., Mercer, J.B. & Jessen, C. (1980). Thermosensitivity of the extrahypothalamic brain stem in conscious goats. *Brain Res.* *188*, 383-397.

Shibolet, S., Lancaster, M.C. & Danon, Y. (1976). Heat stroke: A review. *Aviat. Space Environ.Med.* *47*, 280-301.

Simon, E., Rautenberg, W., Thauer, R. & Iriki, M. (1964). Die Auslosung von Kaltezittern durch loakale Kuhlung im Wirbelkanal. *Pflügers Arch.* *281*, 309-331.

Wyss, C.R., Brengelmann, G.L., Johnson, J.M., Rowell, L.B. & Niederberger, M. (1974). Control of skin blood flow, sweating, and heart rate: role of skin versus core temperature. *J. Appl. Physiol.* *36*, 726-733.

Wyss, C.R., Brengelmann, G.L., Johnson, J.M., Rowell, L.B. & Silverstein, D. (1975). Altered control of skin blood flow at high skin and core temperatures. *J. Appl. Physiol.* *38*, 839-845.

22
Induced Heat Stroke: A Model in Sheep*

M. Khogali
Ghalib Elkhatib
Moneim Attia
M. K. Y. Mustafa
K. Gumaa
Adel Nasr El-Din
Mohamed S. Al-Adnani

Departments of Community Medicine, Biochemistry,
Physiology, Pharmacology and Pathology,
Faculty of Medicine,
Kuwait University, Kuwait

Heat stroke death was induced in 10 Arab and Merino sheep exposed to combined heat and work stress. Two sheep were completely defleeced by oral cyclophosphamide. Rectal temperature and skin temperature on chest, back, thigh, lower leg, testicles and ear were recorded at regular intervals. Blood pressure was measured in the carotid artery. Monitoring of serum enzymes, electrolytes, metabolites, blood gases and pH was undertaken at regular intervals. Post-mortem and histological examinations were performed on different organs.

The results indicated that a critical core temperature of 43.5°C was necessary before heat stroke death. The two defleeced sheep took a markedly longer time, viz., 390 min to reach heat stroke stage as compared with 160-220 min for the fleeced sheep. At heat stroke conditions, concentrations of pyruvate, lactate, creatinine and potassium together with concentrations of glucose, lactate dehydrogenase, creatine phosphokinase and aspartate amino transferase were significantly elevated in all sheep. Concentrations of globulin,

This work was supported by Research Council Grant No. MC 009, Kuwait University and Kuwait Foundation for Advancement of Science.

253

alanine amino transferase, urea and sodium at heat stroke
conditions were not statistically different from levels at
basal conditions. Shortly before heat stroke death, panting
slowed down to a minimum or stopped; skin temperature on the
ear and lower leg decreased and the animal went into convuls-
ions, shivering, coma and eventual heat stroke death. The
results indicate that the sheep is a good model for simulation
of heat stroke in man.

Heat stroke is a complex clinical picture and a medical
emergency with a mortality of up to 80% (Sprung, 1979). The
yearly pilgrimage to Makkah in Saudi Arabia claims many heat
stroke victims when it occurs in summer. Using the Makkah
Body Cooling Unit (MBCU) which was specially designed for the
emergency treatment of heat stroke patients (Weiner & Khogali,
1980), 1500 cases have been successfully treated during a 4-
year period since 1979 (Khogali & Al Khawashki, 1981).
Patients were brought to heat stroke treatment centres with
rectal temperatures of 40.5-45.6°C. Patients arrived in coma,
convulsions and with a hot dry skin. Although the management
of heat stroke patients on the MBCU reduced mortality to a
minimum, it was beset with many problems including metabolic
acidosis, hypokalaemia, hypoxia and bleeding diathesis
(Khogali *et al.*, 1982).

The complex clinical picture associated with heat stroke,
prompted the performance of experiments in the hyperpyrexic
range on different animals (Shibolet *et al.*, 1976). Dogs and
rats were used as heat stroke models to simulate some aspects
of the temperature regulatory system in man (Shapiro *et al.*,
1973, Hubbard *et al.*, 1977). In the present study, however,
an attempt is being made to use sheep as an experimental
model. The purpose is to examine firstly the feasibility of
using sheep to simulate some aspects of the heat stroke episode in man and secondly, to supplement the clinical data
collected during the Makkah pilgrimages and throw some light
on the problems associated with the pathophysiology and
management of heat stroke.

MATERIALS AND METHODS

A. Animals. Ten Arabian and Merino sheep (Table I), with
about 50 mm fleece depth were trained to run on a treadmill.
Approximately 10 days before heat exposure, the right common
carotid artery in each animal was surgically dislodged and

placed on a plate fixed lateral to sternothyroid muscles
(Hales & Webster, 1967), while the sheep was under general
anaesthesia using Halothane in oxygen. Antiobiotic
(Depomycin) was given by intramuscular injection post-
operatively. Two sheep were defleeced by orally administering cyclophosphamide (30 mg/kg body weight) some ten days
before exposure to heat.

B. Temperatures. Initial measurements of core and skin
temperatures were made while the sheep stood quietly in a
trolley in the laboratory prior to each heat exposure.
Rectal (T_{re}), oesophageal (T_{oes}) and skin temperatures on the
chest, back, lower leg, ear and testicles were monitored
using thermistors and digital recorders (Digitron, model
3755). The rectal sensor was inserted 10 cm and the oesophageal thermistor was situated at the heart level. Skin temperature sensors were firmly stuck to the shaved skin with a
strong adhesive.

C. Investigations. During the pre-exposure period two blood
samples were drawn. The first was drawn from the jugular vein
for biochemical analysis of electrolytes, metabolites and
enzymes. Another sample was drawn from the carotid artery
into heparinized syringes for the measurement of arterial
oxygen tension (PO_2), carbon dioxide tension (PCO_2) and pH,
using a blood gas analyzer (Radiometer, Copenhagen, type

TABLE I. Data of Arabian and Merino Sheep

No.	Breed	Sex	Age (months)	Weight (kg)
1	Arabian	M	21	35.0
2	Arabian	M	15	36.5
3	Arabian	M	23	38.0
4	Arabian	M	21	39.0
5	Merino	M	27	30.7
6	Merino	F	24	24.5
7	Merino	F	19	27.4
8	Merino	M	27	26.0
9	Merino	F	27	29.1
10	Merino	F	24	32.0

M, male; F, female

BMS3MK2).   Arterial blood measurements were made at 37.0°C
and corrected to body temperature.   Arterial blood pressure
was recorded by means of a saline-filled carotid catheter,
strain-gauge transducer (Bell and Howell model 4-422-0001)
and a pen recorder.

The sheep was then moved into a climatic chamber, with a
dry bulb temperature of 42-45°C, 40-50% relative humidity and
0.2 m/s air speed.   The sheep was passively exposed to heat
for 120 min while standing on a treadmill and then made to
perform physical work on the treadmill set at 4.5 km/h and an
elevation 4.5%.   Core and skin temperatures were recorded
every 15 min during passive heating and every 5 min thereafter until heat stroke death was observed.   Arterial blood
samples were drawn every 30 min.   Arterial blood pressure was
continuously recorded.   Jugular vein blood was drawn every 60
min.   Samples of venous blood were analysed by standard laboratory methods for the metabolites pyruvate, lactate,
glucose, urea and creatinine, for the enzymes gamma-glutamyl
transferase (γGT), lactate dehydrogenase (LDH), creatine
phosphokinase (CPK), aspartate amino transferase (AST) and
alanine aminotransferase (ALT) and for the electrolytes
sodium and potassium.   The dead sheep was dissected to extract
the heart, brain, liver, kidneys and lungs for histopathological examination.

RESULTS

The results of only eight sheep are reported here.   The
results of the other two were excluded because one had a
septic wound and the other bled profusely.   Both conditions
(infection and bleeding) influenced the outcome of rapid heat
stroke;   the two will be discussed elsewhere.

A.   *Core and Skin Temperature*.   Rectal temperatures of eight
sheep, two of which were defleeced, and points of heat stroke
death are shown in Fig. 1.   During the 2 h period of passive
heat exposure all animals showed an increase in core temperature of 1-2°C.   Following the onset of exercise, core
temperature of the six fleeced sheep showed a sharp increase
averaging 1°C every 30 min.   In the two defleeced sheep a
steep rise in core temperature was observed some 75 min after
commencing exercise in the hot environment.   The total duration of exposure necessary to induce heat stroke was 160-220
min in the 6 fleeced sheep and 390 min in the two defleeced

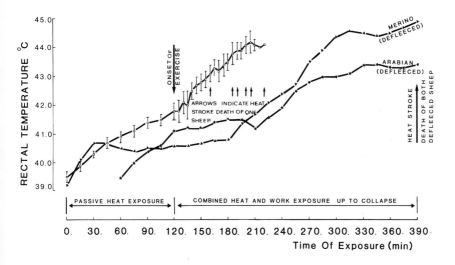

Fig. 1. *Mean ± s.e. rectal temperature of 6 sheep plotted against time of exposure to heat and work. Rectal temperature of 2 defleeced sheep is plotted on same figure for comparison.*

sheep (Fig. 1). Heat stroke death was preceded by coma, convulsions, progressive decrease with eventual stoppage of open-mouth panting, extremity skin temperature decrease and shivering in all sheep. Core temperature at which heat stroke death was observed ranged from 43.7 to 44.9°C.

The course of skin temperatures and core temperature was compared throughout the passive and work periods of exposure (Fig. 2).

During passive heat exposure, skin temperatures on the trunk increased slowly, but those of the lower leg and ear increased sharply on admission to the chamber. This sharp increase indicates an increase in blood flow in the arterio-venous anastomoses to meet demands of heat dissipation. Both skin and core temperatures increased at a higher rate when exercise was imposed on the animal. An increase in panting frequency was recorded with the onset of exercise and open-mouth panting occurred at rectal temperatures above 42.0°C

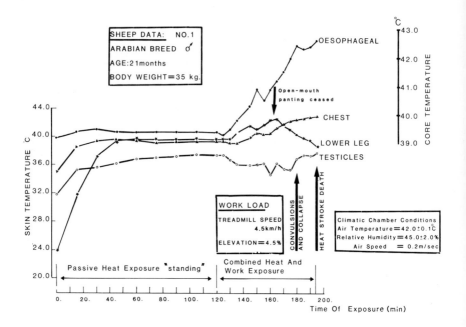

Fig. 2. Oesophageal and skin temperature on chest, lower leg and testicles of a Merino sheep subjected to heat and work stress, plotted against time of exposure.

and increased continually in frequency with excessive thick salivation. A point was reached when open-mouth panting started markedly and progressively to decrease in frequency and stopped at core temperatures of 43.6°C and above. At the same point in time skin temperature on the ear and lower leg started to drop at a slow progressive rate, indicating a decreased blood flow to the extremities.

B. *Biochemistry of Blood.* Comparison of basal and heat stroke concentrations for some of the biochemical data obtained from blood analysis is shown in Figs 3a and 3b.

Heat stroke concentrations of pyruvate, lactate, creatinine and potassium (Fig. 3a) together with heat stroke concentrations of glucose, LDH, CPK and AST (Fig. 3b) were significantly higher than the corresponding basal concentrations in all sheep. Concentrations of globulin, ALT, urea and sodium at heat stroke conditions were not statistically different from the levels at pre-exposure rest.

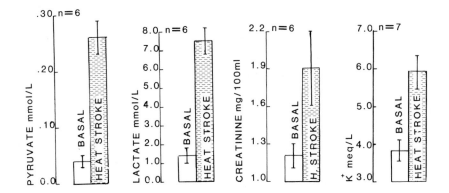

Fig. 3a.   Basal and heat stroke concentrations in venous blood of pyruvate, lactate, creatinine and potassium in 6 sheep.

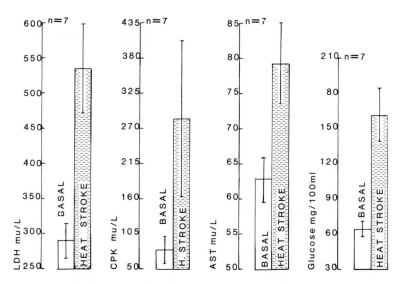

Fig. 3b.   Basal and heat stroke concentrations in venous blood of lactate dehydrogenase (LDH), creatine phosphokinase (CPK), aspartate amino transferase (AST) and glucose in 7 sheep.

C. Blood gases and pH. Oxygen tension in arterial blood showed a progressive increase with the time of exposure to heat and work. Pre-exposure oxygen tension was 90 Torr which increased to 140 Torr at the point of collapse and eventual heat stroke death. Carbon dioxide tension showed the opposite tendency by decreasing from a rest level of 30 to 10 Torr at the point of collapse. The pH increased from 7.4 to a peak of 7.7 but then fell to around 7.5 or even lower before heat stroke death.

D. Histopathology. Histological examination of the extracted organs showed severe congestion of most serosal surfaces and of the organs; only the lungs were almost always involved. The degree of congestion was variable among different sheep. Congestion in brain was rarely seen. Accumulation of white blood cells and fibrin was found in the heart with some suggestion of small focal areas of necrosis.

DISCUSSION

The biochemical and physiological results, as well as the clinical observations documented in the present study were almost identical to our observations in humans. The start of the failure of the temperature regulatory system in our sheep occurred at high rectal temperatures above 43.5°C and was accompanied by convulsions, and decreased panting rates before coma ensued. This was preceded by a significant rise in the levels of metabolism, enzymes and potassium.

The sheep went through a stage of progressively increasing respiratory alkalosis proportional to the increase in panting frequency and when the latter decreased, shortly before collapse, metabolic acidosis predominated. Respiratory alkalosis is probably the cause of a depressed CNS blood flow in sheep (Hales, 1983). The morbid anatomy points to the consequences on the lung, which was the only organ regularly involved in all sheep.

Although evaporative cooling is the major heat loss mechanism in both man and sheep, sweating and panting are different mechanisms. It might be argued that the sheep does not simulate man in this respect. However a very careful comparison of a number of panting and non-panting animals with man, reveals a striking similarity in the cardiovascular (including skin blood flow) response pattern to heat, of sheep and man (Hales, 1983). The present study gives further evidence of this similarity. Moreover we have found that both Arab and Merino sheep are quite amenable to experimental procedures. Also, experiments on sheep induced to the critical temperature of 43.5°C and then cooled on the MBCU gave further evidence of likeness to man (Mustafa *et al.*, 1983). In conclusion, our studies indicate that sheep are a very good model for simulation of heat stroke in man.

REFERENCES

Hales, J.R.S. (1983). Circulatory consequences of hyper-
    thermia: An animal model for studies of heat stroke. *In*
    "Heat Stroke and Temperature Regulation" (M. Khogali &
    J.R.S. Hales, eds), pp. 223-240. Academic Press, Sydney.
Hales, J.R.S. & Webster, M.E.D. (1967). Respiratory function
    during thermal tachypnoea in sheep. *J. Physiol.*
    190, 241-260.
Hubbard, R.W., Bowers, W.D. & Mathew, W.T. (1977). Rat model
    of acute heat stroke mortality. *J. Appl. Physiol.*
    42, 809-816.
Khogali, M. & Al Khawashki, M.I. (1981). Heat stroke during
    the Makkah Pilgrimage. *Saudi Med. J. 2*, 85-93.
Khogali, M., Mustafa, M.K.Y. & Gumaa, K. (1982). Management
    of heatstroke. *Lancet ii*, 1225.
Mustafa, M.K.Y., Khogali, M., Elkhatib, G., Attia, M., Gumaa,
    K., Nasr El-Din, A., Mahmoud, N.A. & Al-Adnani, M.S.
    (1983). Sequential pathophysiological changes during
    heat stress and cooling in sheep. *In* "Thermal Physiology"
    (J.R.S. Hales, ed.)(in press). Raven Press, New York.
Shapiro, Y., Rosenthal, T. & Sohar, E. (1973). Experimental
    heat stroke : A model in dogs. *Arch. Intern. Med.*
    131, 688-692.
Shibolet, S., Lancaster, M.C. & Danon, Y. (1976). Heat
    stroke : A review. *Aviat. Space and Environ. Med.*
    47, 280-301.
Sprung, C.L. (1979). Haemodynamic alterations of heat stroke
    in the elderly. *Chest 75*, 362-366.
Weiner, J.S. & Khogali, M. (1980). A physiological body
    cooling unit for treatment of heatstroke. *Lancet i*,
    507-509.

23
Interaction of Hyperthermia and Blood Vessels

A. Elkhawad
M. Khogali[1]
O. Thulesius
S. M. Hamdan
O. Hassan

Departments of Pharmacology & Community Medicine[1],
Faculty of Medicine,
Kuwait University, Kuwait

This investigation was carried out to determine if alpha-adrenoceptor mechanisms or 5-hydroxytryptamine (5-HT) contribute to vasconstriction of the sheep renal artery during high body temperatures. Moreover, blood vessels obtained from sheep after death from experimentally-induced heat stroke were tested for reactivity to various pharmacological agents. In addition, the effect on the renal artery vasoconstriction of a specific 5-HT$_2$-receptor blocker, ketanserin, was studied.

It was shown that, contractile responses of the renal arteries to the specific α_1-adrenoceptor stimulant phenylephrine (Phe) were unchanged at 37, 39, 41, 42 and 44°C. Contractions elicited by noradrenaline (NA) at 44°C were slightly more pronounced than at low temperatures (lower ED$_{50}$-value). This can be attributed to a desensitization of β_2-receptors mediating vasodilatation.

The contractile responses of renal, mesenteric, muscular and skin arteries were attenuated after heat stroke. The most pronounced reductions were observed as the result of stimulation by 5-HT and NA in the renal and mesenteric arteries respectively. This may be due to a desensitization of these vessels to these transmitters after massive in vivo stimulation during heat stress, implicating 5-HT and noradrenergic mechanisms of vasoconstriction.

Finally, ketanserin proved to be a very effective blocker of contraction elicited by 5-HT in the renal artery.

The normal response to heat stress is vasodilatation of cutaneous blood vessels and a compensatory vasoconstriction of the splanchnic and renal vascular beds. In hyperthermic animals and man the initial vasodilatation in skin vessels is followed by vasoconstriction (Hales, 1983). It has been shown that this vasomotor activity is mediated via α-adrenergic activity in some vascular beds (Walther et al., 1970; Kullmann et al., 1970; Conradt et al., 1975; Simon & Riedel, 1975; Kim et al., 1979; Proppe, 1980; Hales et al., 1982). Another alternative could be an increased activity of free renin-angiotensin system which has been reported for the baboon (Eisman & Rowell, 1977). In men, however, the renin-angiotensin system does not seem to play a major role (Escourrou et al., 1982). In a previous study we were able to show that there was an increased responsiveness of the sheep renal artery to 5-hydroxytryptamine (5-HT) at increasing temperatures (Elkhawad et al., 1983).

The purpose of the present study was to investigate the response pattern of the renal artery to α-adrenergic stimulation at different temperatures. Moreover, we studied the responses of blood vessels from different vascular beds obtained after experimental heat stroke in the sheep. In addition some data are presented on the effect of the selective 5-HT$_2$-receptor blocker, ketanserin.

MATERIALS AND METHODS

Experiments were performed on ring preparations of the isolated sheep renal artery equilibrated for 1 h under an initial tension of 2.0 g in Krebs Henseleit solution at temperatures of 37, 39, 41, 42 and 44°C, gassed with 95% O_2 and 5% CO_2. Isometric tension was measured and displayed on a pen recorder. Blood vessels obtained from sheep immediately after death from heat stroke were put into Krebs Henseleit solution. These blood vessels included renal, mesenteric, ear, cutaneous abdominal and saphenous arteries. The method for induction of heat stroke is described by Khogali et al. (1983). Cumulative dose-response curves were constructed for NA and 5-HT.

All drugs were dissolved in Krebs Henseleit solution: noradrenaline bitartrate (Sigma), phenylephrine HCl (Phe, Sigma), 5-hydroxytryptamine (Sigma), ketanserin (Janssen Pharmaceutica). Drug concentrations represent the final bath concentration.

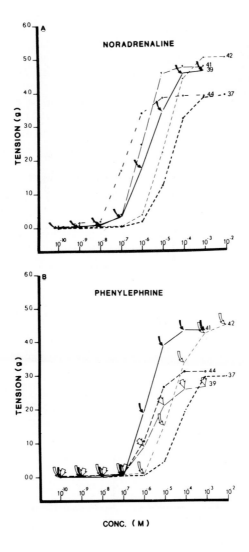

Fig. 1.   Cumulative concentration-response curves at different bath temperatures: (A) for noradrenaline on isolated sheep renal artery.   A significant change was only observed at 44°C (decreased ED_{50}).   (B) for phenylephrine on isolated sheep renal artery.   No significant change was observed. Curves are the mean of several experiments at each temperature, 37 (n = 6), 39 (n = 8), 41 (n = 7), 42 (n = 7) and 44°C (n = 7).

RESULTS

Dose-response curves were obtained after stimulation with NA and Phe at 37, 39, 41, 42 and 44°C. The results are given in Table I and Fig. 1. It is apparent that there are only very few significant temperature-related changes in the dose-response curves: a reduced ED_{50} and ED_{100} were seen for NA stimulation at 44°C, whereas the corresponding values for Phe were unchanged.

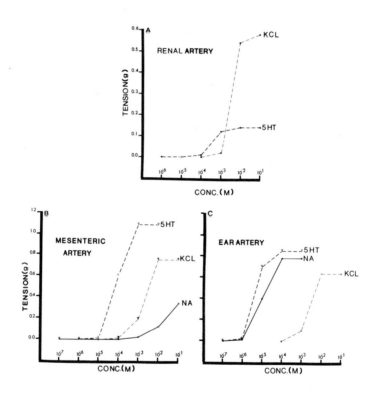

Fig. 2.   Cumulative concentration response curves performed on blood vessels obtained from sheep after death from experimental heat stroke. (A) Responses for potassium chloride and 5-HT on isolated sheep renal artery. (B) Responses for 5-HT, potassium chloride and NA on isolated mesenteric artery. (C) Responses for 5-HT, potassium chloride and NA on isolated ear artery. All experiments were performed at 37°C (n = 8).

Temperature	37°C (6)	39°C (8)	41°C (7)	42°C (7)	44°C (7)
PHENYLEPHRINE					
Threshold	$2 \times 10^{-6} \pm 1.5 \times 10^{-6}$	$6.3 \times 10^{-7} \pm 3.5 \times 10^{-7}$	$4.3 \times 10^{-7} \pm 1.7 \times 10^{-7}$	$7.3 \times 10^{-7} \pm 3.8 \times 10^{-7}$	$1.36 \times 10^{-6} \pm 8.5 \times 10^{-7}$
ED_{50}	$1 \times 10^{-4} \pm 4.7 \times 10^{-5}$	$2.7 \times 10^{-5} \pm 6.1 \times 10^{-6}$	$2.1 \times 10^{-5} \pm 5.1 \times 10^{-6}$	$1.8 \times 10^{-4} \pm 1.7 \times 10^{-4}$	$4.7 \times 10^{-6} \pm 2.5 \times 10^{-6}$
ED_{100}	$9.4 \times 10^{-4} \pm 2.7 \times 10^{-4}$	$5.1 \times 10^{-4} \pm 14 \times 10^{-4}$	$9.4 \times 10^{-4} \pm 1.6 \times 10^{-4}$	$1.2 \times 10^{-3} \pm 2.6 \times 10^{-4}$	$2.5 \times 10^{-4} \pm 1.1 \times 10^{-4}$ **
Maximal response (mg)	2988.3 ± 897.9	2977.5 ± 461.53	4528.8 ± 354.4	5326.6 ± 841.12	2880 ± 515.47
NORADRENALINE					
Threshold	$3.7 \times 10^{-7} \pm 1.7 \times 10^{-7}$	$3.3 \times 10^{-7} \pm 2.2 \times 10^{-7}$	$5.5 \times 10^{-7} \pm 1.5 \times 10^{-7}$	$1.2 \times 10^{-7} \pm 1 \times 10^{-7}$	$2.3 \times 10^{-7} \pm 3.8 \times 10^{-7}$
ED_{50}	$2.4 \times 10^{-5} \pm 5.2 \times 10^{-6}$	$1.8 \times 10^{-5} \pm 6.9 \times 10^{-6}$	$1.2 \times 10^{-5} \pm 2.5 \times 10^{-6}$	$1.2 \times 10^{-5} \pm 2.5 \times 10^{-6}$	$5.6 \times 10^{-6} \pm 2.7 \times 10^{-6}$ *
ED_{100}	$6.8 \times 10^{-4} \pm 9.1 \times 10^{-5}$	$4 \times 10^{-4} \pm 8.6 \times 10^{-5}$	$7.2 \times 10^{-4} \pm 7.6 \times 10^{-5}$	$7.4 \times 10^{-4} \pm 1.7 \times 10^{-4}$	$4.8 \times 10^{-4} \pm 1.9 \times 10^{-4}$ **
Maximal response (mg)	3930 ± 582.26	4366.67 ± 664.9	4922 ± 482.7	5745 ± 831.7	3425 ± 319.2

* $P < 0.02$; ** $P < 0.01$

TABLE I. Threshold, ED_{50}, ED_{100} and maximal response to phenylephrine and noradrenaline at different bath temperatures. Mean values (molar concentrations) ± S.E. Number of experiments in brackets.

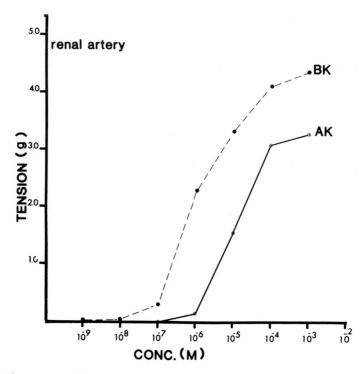

Fig. 3.   Effect of ketanserin on the contractile response to 5-HT.   Concentration-response curves for 5-HT were constructed before (BK) and after (AK) the application of ketanserin 5 × 10⁻¹⁰ M.   The response curves represent the mean of 8 determinations for each experiment.

Various sheep arteries obtained from animals which succumbed to heat stroke were tested *in vitro*:   these included renal, mesenteric, ear, cutaneous abdominal and saphenous. The results are depicted in Fig. 2A, B and C.   In all preparations smooth muscle contraction could be elicited, but at an attenuated level with a decreased maximal tension and an increased threshold.   In the renal artery the previously marked response to 5-HT was considerably reduced, whereas contraction elicited with high molar potassium chloride was better maintained (Fig. 2A).   In the mesenteric artery the response to 5-HT was better preserved and the NA-induced contraction was lowest (Fig. 2B).   In the ear artery both the responses to 5-HT and NA were similar (Fig. 2C).   In the cutaneous abdominal and saphenous arteries, responses were best elicited with 5-HT.

Dose-response curves to 5-HT in the renal artery were repeated after administration of the selective $5-HT_2$-blocker, ketanserin, at 5×10^{-10} M. In these experiments we observed a rightward shift of the dose-response curve (Fig. 3).

DISCUSSION

Alpha-adrenergic stimulation of sheep renal arteries with NA and Phe gave identical contractile responses at temperatures of $37-42°C$. At $44°C$, however, the ED_{50} value for NA was reduced, which means that contraction was more easily achieved. This difference cannot be explained on the basis of an altered α-receptor affinity, since the response to Phe was not changed. NA elicits a dual action by activating both α_1 and β_2 receptors, the latter mediating vasodilatation. The resulting contraction is always a combination of the two opposing effects and therefore, heating might preferentially have down regulated β_2-adrenoceptor function. It has been shown that in isolated ileum and atrium of the rat there is a decreased affinity for beta sympathomimetic drugs (Reinhardt *et al.*, 1973).

The marked decline in contractility seen in the renal artery obtained after heat stroke is difficult to explain, but may have been related to a desensitization after massive stimulation during heat stroke. A definite answer to these propositions cannot be given with the present data, but further experiments with receptor ligand binding studies could be helpful. Moreover the balance of α- and β-receptor activity has to be assessed with additional studies using β-adrenoceptor blockers. The reduced responsiveness of the mesenteric artery to NA could be related to a similar mechanism of receptor down regulation after massive adrenoceptor stimulation.

Recent experience with ketanserin in the treatment of malignant hyperthermia in pigs showed a very favourable response on the temperature reaction and had a curative effect on the disturbance (Ooms & Verheyen, 1982). This is an interesting observation which links 5-HT to another pathological hyperpyrectic disorder in which the role of 5-HT has not been clearly resolved. A peripheral mechanism may be involved since ketanserin is not supposed to have any central actions (Awouters *et al.*, 1982).

In conclusion our data shows that the renal artery does not seem to be sensitized to α-adrenoceptor stimulation but increased sympathetic stimulation could have been operative in mesenteric blood vessels during heat stroke and an increased 5-HT stimulation in renal arteries.

Acknowledgement

We are grateful to Janssen Pharmaceutica for the generous gift of ketanserin, and Kuwait University for financial support.

REFERENCES

Awouters, F., Leysen, J.E., De Clerck, F. & Van Nueten, J.M. (1982). General pharmacological profile of Ketanserin (R.41 468), a selective 5-HT$_2$ receptor antagonist. *In* "5-Hydroxytryptamine in Peripheral Reactions" (F. De Clerck & P. Vanhoutte, eds), pp. 193-198. Raven Press, New York.

Conradt, M., Kullmann, R., Matsuzaji, T. & Simon, E. (1975). Arterial baroreceptor function in differential cardio-vascular adjustments induced by central thermal stimulation. *Basic Res. Cardiol. 70*, 10-28.

Eisman, M.M. & Rowell, L.B. (1977). Renal vascular response to heat stress in baboons - role of renin-angiotensin. *J. Appl. Physiol. 43*, 739-746.

Elkhawad, A.O., Khogali, M. & Thulesius, O. (1983). Vascular effects of hyperthermia on isolated blood vessels. *Gen. Pharmacol. 14*, 69.

Escourrou, P., Freund, P.R., Rowell, L.B. & Johnson, D.G. (1982). Splanchnic vasoconstriction in heat stressed men: role of renin-angiotensin system. *J. Appl. Physiol. 52*, 1438-1443.

Hales, J.R.S. (1983). Circulatory consequences of hyper-thermia: an animal model for studies of heat stroke. *In* "Heat Stroke and Temperature Regulation" (M. Khogali & J.R.S. Hales, eds), pp.   -   . Academic Press, Sydney.

Hales, J.R.S., Foldes, A., Fawcett, A.A. & King, R.B. (1982). The role of adrenergic mechanisms in thermoregulatory control of blood flow through capillaries and arterio-venous anastomoses in the sheep hind limb. *Pflügers Arch. 395*, 93-98.

Khogali, M., Elkhatib, G., Attia, M., Mustafa, M.K.Y., Gumaa, K., Nasralla, A. & Al-Adnani, M.S. (1983). Induced heat stroke: a model in sheep. *In* "Heat Stroke and Temperature Regulation" (M. Khogali & J.R.S. Hales, eds), pp. 253-261. Academic Press, Sydney.

Kim, Y.D., Lake, C.R., Lees, D.E., Schuette, W.H., Bull, J.M., Weise, V. & Kopin, L.J. (1979). Hemodynamic and plasma catecholamine response to hyperthermic cancer therapy in humans. *Am. J. Physiol.* *237*, H570-574.

Kullmann, R., Schönung, W. & Simon, E. (1970). Antagonistic changes of blood flow and sympathetic activity in different vascular beds following central thermal stimulation. I. Blood flow in skin, muscle and intestine during spinal cord heating and cooling in anaesthetized dogs. *Pflügers Arch.* *319*, 146-161.

Ooms, L.A.A. & Verheyen, A.K. (1982). Malignant hyperthermia: etiology, pathophysiology and prevention. *In* "5-Hydroxytryptamine in Peripheral Reactions" (F. De Clerck & P. Vanhoutte, eds), pp. 129-140. Raven Press, New York.

Proppe, D.W. (1980). α-Adrenergic control of intestinal circulation in heat-stressed baboons. *J. Appl. Physiol.* *48*, 759-764.

Reinhardt, D., Wagner, J. & Schumann, H.J. (1973). Changes of the β-receptor binding sites of the rabbit ileum under the influence of high temperature. *Experientia* *29*, 830-832.

Simon, E. & Riedel, W. (1975). Diversity of regional sympathetic outflow in integrative cardiovascular control: patterns and mechanisms. *Brain Res.* *87*, 323-333.

Walther, O.E., Iriki, M. & Simon, E. (1970). Antagonistic changes of blood flow and sympathetic activity in different vascular beds following central thermal stimulation. II. Cutaneous and visceral sympathetic activity during spinal cord heating and cooling in anaesthetized rabbits and cats. *Pflügers Arch.* *319*, 162-184.

24
Contribution of Air Humidity and Heat Radiation to Heat Stress due to Elevated Air Temperature

Hans Gerd Wenzel

Institute of Work Physiology
University of Dortmund
Dortmund, Germany

*Heat stress, which causes increases in body temperature
and possibly heat-related illnesses may be due to elevation
in air temperature. However, other thermal factors also play
a significant role, but their contribution to heat stress is
only partially known.*

*In some 500 climatic chamber experiments, 11 lightly-
clad, young healthy men were exposed to systematically varied
combinations of ambient temperature (total range 15°C to
57°C) and relative humidity (6% to 99%) under different work
loads, usually lasting 4 h. On the basis of changes in body
temperatures, heart rate and sweat loss, physiologically
equivalent temperature-humidity combinations were derived;
these were compared with some indices of heat stress.*

*In 35 similar experiments, 2 subjects were exposed to
globe temperatures (25°C to 50°C) which were produced either
by equal air and radiant temperature, or by stepwise
increases in mean radiant temperature (maximum 89°C) combined
with stepwise reductions in air temperature (minimum +5°C).
Equal globe temperature did not cause equal physiological
effects. Decreasing the air temperature while increasing the
radiant temperature caused decreased sweat loss, heart rate
and rectal temperature. The globe thermometer does not seem
to be a suitable integrating instrument, since it did not
respond to changes in air and radiant temperature in the same
way as the human body did.*

In order to prevent heat-related diseases it is important
to be able to predict whether or not a given hot environment
might overload the body's thermoregulatory efficiency. Such
predictions are at least approximately correct in those cases
in which data on health risks are collected under conditions
of heat stress that are comparable to those conditions to be
assessed. Apart from a number of individual factors, however,
various climatic factors may combine to play a significant
role, which may be quite different from one heat stress
situation to another. In order to extrapolate data collected
under certain conditions to predict the effect of combinations that have not yet been observed in detail, knowledge
is required about what combinations of climatic, and perhaps
non-climatic, factors yield equal effects. Unfortunately,
opinions about this question are divided, as relatively few
systematic investigations on this problem have been done
until now.

We have for a number of years, carried out laboratory
studies in climatic chambers in order to determine combined
effects. We first studied the effect of elevated ambient
temperature and air humidity on very lightly-clad subjects
who performed different muscular work lasting several hours.
The results obtained from the first three subjects have been
described previously (Wenzel, 1978). In order to learn more
about interindividual differences of physiological responses
which were found to be of great importance, the experiments
were continued with additional subjects. The first part of
the present report is a survey of how heat stress due to
elevation in ambient temperature is increased by an additional elevation in air humidity, which hinders the essential
thermoregulatory possibility of man to keep thermal equili-
brium by evaporative heat loss. These studies were carried
out together with R. Ilmarinen and C. Piekarski.

The second part of the report describes experiments in
which subjects were exposed to thermal conditions with
unequal air and radiant temperatures. Although heat
radiation plays a particular role as a heat stress factor
under many conditions in daily life, few observations of its
effect were available until now. These studies were carried
out together with J. Sušnik and B. Kampmann.

CONTRIBUTION OF AIR HUMIDITY

A. Techniques

Eleven healthy young male volunteers participated in some 500 experiments performed in climatic chambers which have been described previously (Wenzel & Stratmann, 1968; Wenzel *et al.*, 1980). Individuals who displayed factors that might predispose them to heat diseases have obviously been excluded from the experiments. The subjects wore shorts, socks and shoes, and were supplied with drinking water *ad lib*. Several series were carried out with each subject. Every series consisted of about 20 to 35 single experiments in which combinations of ambient temperature (total range 15°C to 57°C) and relative humidity (6% to 99%) were varied systematically. Air speed was constant at 0.3 m/s. From one series to the next, men were exposed either while sitting at rest or while walking on a treadmill on the level or uphill. Metabolic rates reached about 350 to 1300 kJ/h. The combinations of variables were chosen in such a way that they were usually tolerable for at least 4 h. All working experiments consisted of 30 min walking periods interrupted by 3 min rest. They started at about 8.30 a.m. after at least 1 h rest in bed under thermally neutral conditions.

Before and during the exposures, rectal temperature and heart rate were measured continuously. Skin surface temperatures, mostly at nine areas, were measured in the first series during the rest periods, together with body weight; in later experiments skin temperatures were measured continuously. Extrarenal water loss (mainly sweat) was calculated from changes in body weight corrected for O_2 uptake, CO_2 loss and fluid intake. Metabolic rate was calculated from indirect calorimetric measurements (Wenzel, 1978) half-way through the 30 min periods mentioned above. Before the experiments the subjects were acclimated to work in heat by daily repeated heavy treadmill work in a warm, humid climate that produced continuous increases in body temperatures and heart rate, but which allowed the highly motivated men to work for at least 2 h. The acclimation period lasted about four weeks. As the subsequent experimental periods extended over several months, reacclimation exposures were made each week, partly during the weekends and partly on the following weekday(s).

B. *Results and Discussion*

The combinations of temperature and humidity in each
series may be seen in Fig. 1, which presents rectal temperatures of one subject at the end of the 2nd hour of 38
exposures. O_2 uptake of the man walking on the level reached
about 0.65 1/min. The symbols indicate the effects of air
humidity at ambient temperatures between 24°C and 50°C. As
shown previously for the first three subjects, similar
diagrams were obtained for skin temperature and heart rate.
In all of the series the relationships between the values of
body temperatures and heart rate, reached after at least one
hour of equal work, were independent of the combinations of
ambient temperature and air humidity used. This observation,
however, did not apply to weight (sweat) loss; these findings agree with other reports (Ilmarinen, 1978; Robinson,

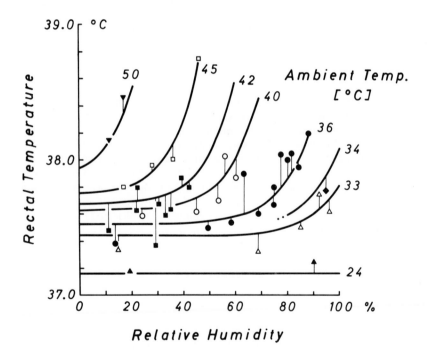

*Fig. 1. Rectal temperatures obtained in one of the
series (38 experiments, Subj. OF).*

1949; Robinson *et al.*, 1945). According to a mathematical procedure proposed by Bronstein & Semendjajew (1971), a single equation was developed for each subject, that described the relationship between body temperatures and heart rate on one hand and metabolic rate, ambient temperature and air humidity on the other hand. Comparisons between measured and calculated values gave correlation coefficients that usually lay between 0.90 and 0.95. As an example, the curves drawn in Fig. 1 are calculated from this equation.

Those combinations of ambient temperature and air humidity leading to equal increases in rectal temperature, mean skin temperature and heart rate in each subject were calculated by means of this equation and plotted as "equivalence lines" in psychrometric charts. Using the procedure previously used for summarizing the individual equivalence lines of the first three subjects, the single lines obtained from the present subjects were combined to "mean equivalence lines". The lower diagram of Fig. 2 shows mean rectal isotherms of nine men derived from all experimental series in which metabolic rate reached about 800 to 900 kJ/h. As shown previously, the slopes of the equivalence lines changed systematically with metabolic rate (Wenzel, 1978). For this reason the results of two out of the eleven subjects were not included in Fig. 2 as these men were not studied at this particular metabolic rate.

The slopes of the rectal isotherms show that the effect of a given increase in air humidity differed in various ranges of climate. The effect was minimal in relatively low air temperatures, and increased gradually with air temperature. This increase was smaller at a low water vapour pressure of the air, and greater at high humidity. These differences correspond to the changes of requirements and possibilities of evaporative cooling of the body.

Considerable differences of individual rectal isotherms were observed. The standard deviations (two examples of which are shown by shaded areas in the upper and lower sections of Fig. 2) decrease as heat stress increases. The rectal isotherms are compared with some indices of heat stress in the upper diagram. The degree of coincidence of various indices with rectal isotherms depended upon the range of climate. The 38.2°C rectal isotherm is fairly well described by the simply calculable values of WD Index (Lind *et al.*, 1957) and WBGT Index (Minard *et al.*, 1957), and it also corresponds well to a constant value of Index of Physiological Effect (Robinson *et al.*, 1945). According to most investigators, rectal temperatures in about this range are still just "safe" or "acceptable" (Wenzel & Piekarski,

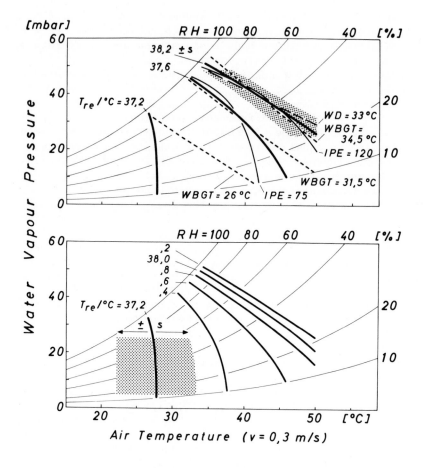

Fig. 2. Rectal isotherms, means and S.D., nine subjects, exposure time three hours. Upper diagram: Comparison with some indices.

1982). Contrary to this result only a partial coincidence was observed in slightly lower climates, corresponding for instance, to the 37.6°C rectal isotherm, while equal values of WBGT (as well as of WD and Effective Temperature Index (Yaglou, 1927), not shown in the diagram) disagreed with rectal isotherms in still lower climatic ranges under the present conditions, overestimating the contribution of air humidity to heat stress.

The results shown in Fig. 2 should not yet be generalized as they were obtained from very lightly-clad subjects who performed medium muscular work. Such conditions may apply to some situations in daily life. The slopes of the equivalence lines, however, depend upon metabolic rate (Wenzel, 1978) as well as upon insulation values of clothing (Ilmarinen, 1978), as shown in a single experimental series on a few test subjects to date. The individual differences of responses observed in the present studies point to the need for further investigations on a greater number of subjects at different activities and wearing different clothing, in order to define more exactly the possibilities for predicting the contribution of air humidity to heat stress.

CONTRIBUTION OF RADIANT HEAT

Although many climates to which man is exposed are characterized by radiant temperatures higher than air temperature, it is not known what combinations of these two thermal factors produce equal effects on man. The Vernon globe thermometer, which was developed several decades ago, is still the instrument that is most frequently used to evaluate such conditions (Vernon & Warner, 1932). It seems to be unknown, however, within which ranges of unequal air and radiant temperature this instrument "reacts" in the same way as does the human body.

A. Techniques

In 35 experiments, most of which lasted 4 h, two healthy young men were exposed to conditions with globe temperatures between 25°C and 50°C, always with 0.5 m/s air speed and low humidity. The experiments were carried out in a special climatic chamber in which air and radiant temperatures could be varied independently of each other within wide ranges (Wenzel *et al.*, 1980). The technical principles of this laboratory have been described previously (Müller & Wenzel, 1957).

The desired globe temperature was produced either by equal air and radiant temperature, or by a stepwise increase in mean radiant temperature (maximum 89°C) combined with a stepwise reduction in air temperature (minimum +5°C). The heat radiation reaching the subjects from various horizontal directions was essentially equal, with differences of not more than 12% from the mean value.

In all the experiments the subjects walked at 4 km/h on the level, with O_2 uptake reaching about 0.7 1/min. The course of the exposures, time of the day, clothing worn by the subjects, their water supply and the physiological measurements were the same as described above in 'Contribution of Air Humidity'.

B. *Results and Discussion*

As examples of the results, data from both subjects at 45°C globe temperature are shown in Fig. 3. The abscissas indicate which combinations of air and radiant temperature were used; these varied from equal air and radiant temperatures (at the left) to 5°C air temperature and 89°C radiant temperature (at the right). The mean radiant temperatures were calculated by the equation of Bedford & Warner (1934). All the combinations produce the same (45°C) globe temperature, however heart rate and rectal temperature at the end of the 3rd hour of work decreased significantly with decreasing air temperature and correspondingly elevated radiant temperature. The same applied to weight loss during the 3rd hour of exposure. Qualitatively the same results were observed in all the series carried out in 25°C, 35°C and 50°C globe temperatures. The decreases of the physiological responses, as the air and radiant temperatures were made less alike, were greater at higher globe temperatures. At least under the present conditions, unequal combinations of both the ambient temperatures that produce a constant globe temperature, produced systematically different physiological responses. Sweat loss, for instance, differed by up to about 300 g/h under the extreme conditions, and heart rate up to about 30 beats/min. This means that the globe thermometer underestimated the effects of decreasing air temperature and considerably overestimated the effects of correspondingly increasing radiant temperature.

It seemed important to carry out the experiments described, as conditions like these are not rare, and because it is frequently assumed that globe temperature reflects the combined effect of various air conditions and heat radiant intensities. Although the present studies did not yet include variations of additional thermal factors such as air speed, air humidity or clothing it is certainly advisable to use globe temperatures cautiously for evaluating conditions like those described above. Further research is surely needed to assess the practical importance of heat radiation as a stress factor.

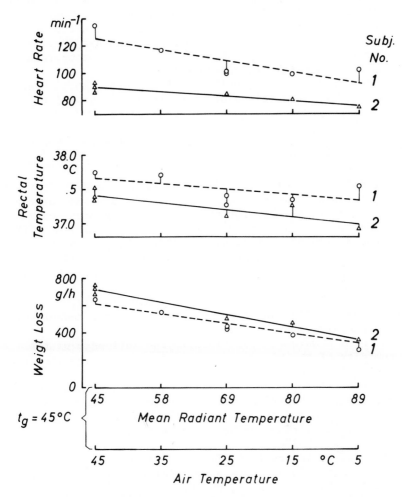

Fig. 3.   Physiological responses of two subjects at 45°C globe temperature at unequal air temperatures and radiant temperatures.

REFERENCES

Bedford, T. & Warner, C.G. (1934).   The globe thermometer in studies of heating and ventilation.   *J. Hyg.*, *Camb.* *34*, 458–473.

Bronstein, I.N. & Semendjajew, K.A. (1971). "Taschenbuch der Mathematik", 11. Auflage, Verlag H. Deutsch, Zürich.

Ilmarinen, R. (1978). Enflüsse verschiedener Bekleidung auf einige physiologische Größen des Menschen bei Körperarbeit in unterschiedlich erhöhter Umgebungstemperatur und Luftfeuchtigkeit. Dissertation Köln.

Lind, A.R., Weiner, J.S., Hellon, R.F., Jones, R.M. & Frazer, D.C. (1957). Reactions of mines-rescue personnel to work in hot environments. National Coal Board, *Med. Res. Memo No. 1.*

Minard, D., Belding, H.S. & Kingston, J.R. (1957). Prevention of heat casualties. *J. Amer. Med. Ass. 165,* 1813-1818.

Müller, E.A. & Wenzel, H.G. (1957). Eine neue Methode zur Herstellung ungleicher Luft- und Strahlungstemperaturen in einer Klimakammer für physiologische Untersuchungen. *Z. Angew. Physiol. einschl. Arbeitsphysiol. 16,* 373-388.

Robinson, S. (1949). Tropics. *In* "Physiology of Heat Regulation and the Science of Clothing", Part II, (L. Newburgh, ed.), pp. 338-351. W.B. Saunders Comp., Philadelphia-London.

Robinson, S., Turrell, E.S. & Gerking, S.D. (1945). Physiologically equivalent conditions of air temperature and humidity. *Am. J. Physiol. 143,* 21-32.

Vernon, H.M. & Warner, C.G. (1932). The influence of the humidity of the air on capacity of work at high temperatures. *J. Hyg., Camb. 32,* 431-463.

Wenzel, H.G. (1978). Heat stress upon undressed man due to different combinations of elevated environmental temperature, air humidity, and metabolic heat production. A critical comparison of heat stress indices. *J. Human Ergol. 7,* 185-206.

Wenzel, H.G. & Piekarski, C. (1982). "Klima und Arbeit", 2. Auflage, (Bayerisches Staatsministerium für Arbeit und Sozialordnung, ed.), München.

Wenzel, H.G. & Stratmann, F. (1968). Technische Erfahrungen über Bau und Betrieb einer Klimakammer für arbeitsphysiologische Untersuchungen am Menschen. *Internat. Z. Angew. Physiol. einschl. Arbeitsphysiol. 25,* 235-278.

Wenzel, H.G., Piekarski, C., Kampmann, B., Andorf, P. & Schulz, P. (1980). Klimasimulation industrieller Arbeitsplätze im Laboratorium. Bericht der 20. Jahrestagung, Deutsche Gesselschaft für Arbeitsmedizin, pp. 577-579. Gentner-Verlag, Stuttgart.

Yaglou, C.P. (1927). Temperature, humidity and air movement in industries. The effective temperature index. *J. Industr. Hyg. 9,* 297-309.

25
Influences of Physical Training, Heat Acclimation and Diet on Temperature Regulation in Man

Otto Appenzeller

Departments of Neurology and Medicine
University of New Mexico
School of Medicine
Albuquerque, New Mexico

Muscular activity places burdens on the thermoregulatory system and on skin vasomotor activity. Increased body temperature due to increased metabolism of exercising muscles is perceived in the central nervous system. A central temperature threshold for vasodilatation in the cutaneous vascular bed exists which allows increased blood flow proportional to the increased temperature. However, when cardiac output falls, cutaneous vasoconstriction occurs to redirect the blood to the heart, but this reduces heat transfer from the core to the surface. When dehydration is present, the threshold for cutaneous vasodilatation shifts to a higher central temperature implying a centrally generated inhibition of the vasodilator drive.

Endurance training does not increase sweating. Sweat onset occurs, however, at lower central temperatures and continuous aerobic activity enhances sweating sensitivity. Heat acclimation in fit and unfit subjects causes lower central temperatures for a given amount of work in hot conditions because thermal loads are dissipated more efficiently. The neurogenic influences achieving shifts in central temperature thresholds with training are not clear.

Diet-induced thermogenesis is deranged in obesity and glucose intake may be associated with decreased capacity of heat dissipation both during exercise and at rest. The effect of meals and the timing of food intake needs to be evaluated in relation to heat stroke.

EXERCISE

During physical activity, there is a demand for increased
blood supply to several tissues, particularly muscle for increased metabolic requirements and skin to transfer excess
heat. This is such an important factor in the determination
of heat tolerance, that it is also considered in this Workshop by Robertshaw (1983) and Hales (1983). Thus, in thermally neutral conditions, the cardiac output increases proportionally to the rate of oxygen uptake during exercise, and the
increase is predominantly diverted to skeletal muscles. However, if the demand for skin perfusion is high because of
excessive thermal loads imposed by contracting muscles or by
high ambient temperatures, the simple relationship between
cardiac output and oxygen uptake may no longer be present.
Cutaneous blood flow is determined largely by body temperature. At the onset of exercise, metabolic heat production
increases at a rate directly proportional to exercise intensity, and it far exceeds the capacity of the skin for heat
dissipation. Because of this imbalance, body temperature
rises until heat loss through the skin by increased blood
flow and sweating equals heat production. When this occurs,
a new steady level of internal body temperature is reached
and this is maintained until further heat loads either change
the level upwards or decreased workloads lower body temperature again. But, the combined circulatory demands of muscles
and skin are also, in part, determined by average skin temperature, which, in turn, closely relates to environmental temperature. Therefore, when exercise intensity is great or prolonged and ambient temperature is high, circulatory demands
are considerably increased. Under such conditions, the heart
must provide sufficient blood both to muscles and skin or
compromise delivery of blood to one or the other tissue. If
heavy exercise occurs in very hot environments, contracting
muscles demand a disproportionately large segment of the
cardiac output which is not available for the skin. This
limits heat dissipation and therefore leads to progressive
hyperthermia which, in turn, limits exercise. If, however,
adequate blood flow to the skin is maintained at the expense
of contracting muscles, anaerobic work ensues and ATP synthesis is decreased along with the ability to maintain con-
tinuing contractions of skeletal muscles. To avoid compromising blood flow to muscles, skin, or any other tissue during
exercise in the heat, cardiac output would have to increase
continuously. However, with increasing exercise intensity in
neutral and hot environments, splanchnic blood flow is

progressively reduced, and if comparisons are made for the same workloads in cool or hot conditions, the blood flow to the splanchnic bed is predominantly affected by hot environmental conditions (see Rowell, 1974). This redistribution of blood might, in part, compensate for the additional demands of the skin, but in extreme conditions of exercise and heat, skin and muscle circulation and homeothermy are jeopardized.

The competition between skin and muscle for blood flow during exercise is further complicated by an increase in cutaneous venous volume. This increase is thought to be due to dilatation of highly compliant skin veins in response to the thermoregulatory drive - heat transfer is enhanced because of decreased velocity of flow and increased heat exchange time. This increased cutaneous venous volume reduces central blood volume and this, in turn, reduces cardiac filling which potentially compromises cardiac output. If a reduction in stroke volume occurs, a compensatory increase in heart rate is necessary to maintain cardiac output (see Rowell, 1974; Nadel, 1977). When near maximal heart rates are attained and cardiac output is falling due to reduced filling pressure, a cutaneous vasoconstriction is invoked by low-pressure baroreceptors (see Hales, 1983), or possibly by the introduction of other factors such as increased circulating levels of catecholamines (see Robertshaw, 1983).

A similar situation may arise when central blood volume decreases during exercise (relatively short or prolonged) due to loss of plasma water to the extravascular compartment. The cause of this is thought to be tissue hyperosmolality, occurring predominantly, if not exclusively, in active muscles. The transcapillary movement of fluid, however, could also result from increased filtration pressure in active muscles. The absolute fluid loss from the intravascular compartment during exercise is most closely related to the intensity and duration of exercise rather than to ambient temperatures. Central blood volume is further compromised by fluid loss through sweating which is directly related to ambient temperatures, intensity of exercise, and its duration. The sweat rate may exceed 1-1.5 ℓ/h during prolonged exertion in hot environments.

In desert heat, short-term exercise can be performed provided heart rate increases to compensate for any reduction in stroke volume due to body fluid loss. However, this is not the case in endurance events where maximum heart rates may be associated with less than maximum oxygen consumption.

HYDRATION

Profound effects of dehydration on body temperature and circulation are noted. Experimentally, core temperatures are higher in dehydrated subjects at a given intensity of work or at rest in a warm environment. Reduced sweating and forearm blood flow are observed during exercise in dehydrated states (Nadel, 1980). The mechanism by which the thermoregulatory system is affected by hydration is controversial. It is not clear whether dehydration changes thermoregulatory activity of central nervous system neurons or whether it decreases the effectiveness of thermoregulatory effector systems in response to thermal loads. For example, impaired cardiac output during exercise while dehydrated, results from its dependence upon initial blood volume. Exercise with a high initial blood volume allows the exercise-induced decrease in plasma volume to cause less circulatory strain than if the same plasma volume were lost with a low initial blood volume; initial hypovolaemia leads to earlier circulatory adjustments. It is, however, not clear whether such adjustments impair the peripheral circulation (in the skin) to maintain skeletal muscle perfusion, nor is it known how cutaneous blood flow is affected if the muscles remain adequately perfused during dehydration-associated exercise. Dehydration increases the temperature threshold for exercise-induced cutaneous vasodilatation, but once vasodilatation occurs, the relationship between central temperature and blood flow is similar to that in normally hydrated exercisers (see Robertshaw, 1983). Therefore, it appears that the central nervous system is responsible for the vasodilator drive because a change in gain of the thermal eye does not occur; heat transfer during exercise through the cutaneous circulatory bed is as responsive per unit of central temperature change in hypovolaemic as in normovolaemic individuals. However, an additional influence of dehydration is to cause the vasconstrictor influence superimposed upon the heat-induced cutaneous vasodilatation to be activated at very much reduced blood flows.
    Experiments on subjects with mild hyperosmolality have shown that central nervous system osmoreceptors are also involved in the control of the peripheral circulation. They are, however, subordinate to the "volume receptors" because the hypovolaemic vasodilatory shift appears without an increase in plasma osmolality.

PHYSICAL TRAINING

Responses to heat and exercise of fit and unfit subjects are different. For example, during cycle ergometer exercise for 20-25 min at 40-70% of maximal aerobic power at different ambient temperatures, fit, but not endurance-trained subjects maintain constant temperature and circulation at all but very heavy levels of exertion in the heat. The oesophageal temperature (an approximation of body core temperature) at the end of heavy exercise averages 38.1°C during performance in comfortable ambient temperatures, but in an ambient temperature of 36°C, it reaches almost 39°C. Cardiac output in steady state exercise is appropriate to oxygen uptake, but significantly higher during moderate exercise in hot ambient conditions in order to deliver the appropriate amount of blood to muscles and skin. Duringheavy exercise in the heat, it becomes more and more difficult to increase cardiac output because heart rates are approaching maximum levels, and therefore stroke volume is limited. Under such conditions, a relative cutaneous vasoconstriction superimposed on the heat-induced vasodilatation may be sufficient to maintain cardiac output as discussed earlier, i.e., a further reduction in stroke volume is prevented and the central circulation is stabilized. The price paid for circulatory stability is decreased heat transfer from the core to the surface and consequently increased body temperature after about 20 min exercise. Therefore, in non-endurance-trained and non-heat-acclimated individuals, circulatory regulation takes precedence over temperature homeostasis.

The importance of peripheral pooling of blood during exercise in a hot environment is clearly demonstrated by the effects of body position (Rowell, 1974). Thus, the cutaneous vasoconstriction which occurs when the stress is severe, is greater in the upright than semi-upright position. If peripheral venous hydrostatic pressure is reduced as in supine exercise, the increased venous return removes the need for cutaneous vasoconstriction, at least with moderate exertion (Roberts & Wenger, 1980). This position-related vasoconstriction seems to be attributable to the action of cardiopulmonary baroreflexes (see Hales, 1983), and if absent in the upright posture at critical levels of exercise in the heat, may lead to a precipitous fall in stroke volume and cardiac output, thereby forcing termination of exercise and possibly leading to collapse.

Observations on trained runners who have not been specifically acclimated to heat, show that they have improved heat tolerance, and therefore exercise hyperthermia which occurs during training at any environmental temperature, may act as an acclimatization method. Thus, highly trained, but heat-unacclimated individuals may not really exist. Reports that training in humid heat further improves heat tolerance during subsequent performance at moderate ambient temperatures have also appeared (Wells *et al.*, 1980). The converse is also true, viz., heat acclimation offers cardiovascular advantages for subsequent work. Moreover, anecdotal claims from the Honolulu marathon, which is always staged in hot and humid conditions, show that Hawaiian runners perform better than those from the mainland. If one assumes that the central temperature is the main drive for an increasing cutaneous blood flow with work, then acclimatized and endurance-trained individuals have less cardiovascular strain during work in the heat.

ENDURANCE TRAINING, HEAT ACCLIMATION AND PLASMA VOLUME

Insights into the mechanisms underlying cardiovascular changes and regulation of body temperature in endurance-trained and heat-acclimated individuals may point to ways of improving performance during heat stress in athletes (Appenzeller, 1982).

Sweating rates are probably not changed by endurance training, but endurance-trained individuals sweat at lower central temperatures and continuous aerobic activity enhances sweating sensitivity. Because heat acclimation in fit and unfit subjects causes lower central temperatures for a given amount of work in hot ambient conditions, a smaller fraction of the cardiac output need go to skin to maintain temperature homeostasis. This might result from increased sweating sensitivity after acclimation. Therefore, heat acclimation decreases cardiovascular strain in hot environments. When external heat loads compromise performance, endurance training offers an added advantage by decreasing cardiovascular and homeothermic regulatory stress. It is clear that heat- and exercise-acclimated individuals dissipate thermal loads more efficiently, and therefore peripheral circulatory demands are decreased, which in turn improves cardiovascular performance and the capacity for continued skeletal muscle perfusion (Wells *et al.*, 1980).

During acclimatization to heat without exercise, there is a 10-25% expansion of plasma volume. An isotonic expansion of the interstitial fluid also occurs. Close parallels are found between increased plasma volume and thermoregulatory and cardiovascular adaptation to heat. Increased plasma volume enhances heat tolerance and improves physical performance during high ambient temperatures by decreasing body core temperature and cardiovascular strain. In addition, endurance training also increases blood volume mainly due to increased plasma volume. Hypervolaemia is, of course, associated with decreased heart rate and increased stroke volume, both at rest and during exercise, and a reduced haematocrit. Some of these factors contribute to an increased maximum oxygen uptake and cardiac output. There are close parallels between thermoregulatory adaptive responses to heat and those in response to endurance training. In both conditions, there is an increased sweat rate at lower body temperatures, decreased heat storage, and a decreased core temperature at given work loads. Plasma volume expansion during heat acclimatization is a thermoregulatory adaptive response. However, most studies in humans use exercise in addition to heat exposure. Detailed investigations suggest that hypervolaemia is associated with a 40% thermal factor and a 60% non-thermal exercise factor. The non-thermal exercise-induced responses are: a two-fold increase in plasma osmotic and vasopressor forces during exercise and a five-fold increase in resting plasma protein, all of which contribute to hypervolaemia. When individuals were exposed to either exercise training or heat, both of which increased their rectal temperatures to the same levels, plasma volume increased more during exercise training in a cooler environment than during testing in a hot environment, supporting the proposition that metabolic factors in addition to external heat loads are important in elevating plasma volume during endurance training in both hot and cold ambient temperatures. Thus, increased metabolism or other exercise triggers are necessary for maximum expansion of plasma volume and peak performance under hot environmental temperatures (Convertino *et al.*, 1980). To maintain heat and exercise-training-induced adaptive responses, heat exposure needs to be maintained intermittently at four-day intervals. The adaptive expansion of plasma volume decreases considerably within one week when no further stresses are applied. The main stimulus to hyperproteinaemia is exercise and not heat exposure. It is not known, however, whether the additional protein is derived entirely from interstitial spaces in the skin or from increased synthesis of decreased degradation.

DIET AND HYPERTHERMIA

The sympathoadrenal system is important in determining
metabolic and cardiovascular activity. It is, therefore, of
crucial significance in the adaptive responses to both heat
and exercise. Recently, it has been shown that the sympathetic nervous system responds to changes in caloric intake.
Caloric restriction decreases and carbohydrate administration
increases sympathetic nervous system activity, including
blood pressure in both animals and man. It has been suggested that insulin could be an important link between the changes
in dietary intake and changes in central sympathetic outflow.
Moreover, stimulation of sympathetic activity by overfeeding
may be contributory in the development and maintenance of
hypertension in those predisposed to this condition. The
well known association of obesity and hypertension is thought
to reflect chronic overfeeding in animals and man, but diet-
induced changes in sympathetic nervous system activity affect
blood pressure even in non-obese subjects.

Diet-induced thermogenesis is deranged in obese animals
and man. It has been suggested that this derangement results
from abnormal dietary regulation of sympathetic activity, but
this has not been generally accepted. A scheme has been proposed which might explain the dietary effects on sympathetic
activity (Fig. 1): Brainstem centres controlling sympathetic
outflow are inhibited by the hypothalamus. When the hypothalamus is destroyed in animals by gold thyroglucose, des-
cending inhibitory influences are removed. In such animals,
fasting increases the activity of this inhibitory pathway
while sucrose feeding reduces it. A decrease in sympathetic
activity occurs in animals fed, but treated with 2-deoxy-
glucose, a non-metabolizable glucose analogue. Based on
these experiments, it is suggested that the effects of fasting and sucrose feeding result from changes in the rate of
insulin-mediated glucose metabolism within some neurones of
the hypothalamus. This connection between dietary intake
(mostly carbohydrate intake) and sympathetic nervous system
activity has important implications for exercise and heat
acclimation and also for obesity and hypertension (Young &
Landsberg, 1982). If training is carried out in a fasting
state, the hypothalamus is activated; brainstem centres are
inhibited, and this, in turn, decreases the sympathetic outflow, presumably decreasing blood pressure and perhaps allow-
ing better heat transfer from the core to the surface. The
converse happens if heat loads are imposed or training is

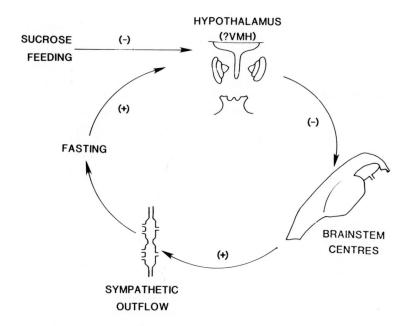

Fig. 1. A hypothetical scheme to explain dietary effects on sympathetic tone. Sucrose inhibits hypothalamic activity; this, in turn, removes the inhibitory effects of the hypothalamus on brainstem autonomic centres which then stimulate sympathetic outflow unchecked. Fasting has the opposite effect by activating the hypothalamus, eventually leading to a decrease in sympathetic tone. (- inhibition; + activation; VMH = ventro-medial hypothalamus)

carried out after carbohydrate administration; this results in an increase in sympathetic nervous system activity.

It is possible, therefore, that heat exposure or training soon after glucose intake is associated with decreased capacity for heat dissipation and proneness to heat stroke while high blood sugar levels persist during exercise and at rest (Young & Landsberg, 1982).

REFERENCES

Appenzeller, O. (1982). "The Autonomic Nervous System" Third revised and enlarged edition. Elsevier Biomedical Press, Amsterdam.

Convertino, V.A., Greenleaf, J.E. & Bernauer, E.M. (1980). Role of thermal and exercise factors in the mechanism of hypervolemia. *J. Appl. Physiol. 48*, 657–664.

Hales, J.R.S. (1983). Circulatory consequences of hyperthermia: An animal model for studies of heat stroke. *In* "Heat Stroke and Temperature Regulation" (M. Khogali & J.R.S. Hales, eds), pp. 223–240. Academic Press, Sydney.

Nadel, E.R. (ed.)(1977). "Problems with Temperature Regulation During Exercise." Academic Press, New York.

Nadel, E.R. (1980). Circulatory and thermal regulation during exercise. *Fed. Proc. 39*, 1491–1497.

Roberts, M.F. & Wenger, B.C. (1980). Control of skin blood flow during exercise by thermal reflexes and baroreflexes. *J. Appl. Physiol. 48*, 717–723.

Robertshaw, D. (1983). Contributing factors to heat stroke. *In* "Heat Stroke and Temperature Regulation" (M. Khogali & J.R.S. Hales, eds), pp. 13–29. Academic Press, Sydney.

Rowell, L.B. (1974). Human cardiovascular adjustments to exercise and thermal stress. *Physiol. Rev. 54*, 75–159.

Wells, C.L., Constable, S.H. & Haan, A.L. (1980). Training and acclimatization: Effects on responses to exercise in a desert environment. *Aviat. Space Environ. Med. 51*, 105–112.

Young, J.B. & Landsberg, L. (1982). Diet-induced changes in sympathetic nervous system activity: Possible implications for obesity and hypertension. *J. Chronic Dis. 35*, 879–886.

26
Prevention of Heat Stroke: Is it Plausible?

M. Khogali

Department of Community Medicine,
Faculty of Medicine, Kuwait University, Kuwait

A. Al-Marzoogi

Ministry of Health, Saudi Arabia

*Heat stroke is a medical emergency of the highest prior-
ity with potential high mortality among healthy young adults
and very high mortality, up to 80%, among the elderly. High
body temperature, decreased consciousness and hot dry skin
constitute the classical triad of the heat stroke syndrome.
Prevention has been sought for years, but is it possible in
all circumstances?*

*Pilgrims come to Makkah from nearly all over the world.
The majority are elderly with concomitant chronic and/or
endemic diseases which complicate the picture. Sanitation,
overcrowding, noise and lack of sleep are major problems. The
Hajj rituals entail strenuous physical exercise.*

*A classical programme of heat stroke prevention should
take into account all human factors, such as acclimatization,
age, build, general health, water and salt intake, clothing,
religious devotion and liability to neglect, or ignorance of
regulations. It should start in the pilgrim's country of
origin and his own language.*

*Our objective is to provide evidence supporting the hypo-
thesis that heat stroke cannot be totally prevented in the
conditions prevailing in the pilgrimage during the hot
seasons. Thus, secondary prevention programmes must be given
more emphasis.*

HEAT STROKE AND
TEMPERATURE REGULATION
ISBN 0 12 406180 X

INTRODUCTION

Heat stroke is a medical emergency of the highest priority with potential high mortality of up to 80% (Bartley, 1977; Knochel, 1974). Relatively healthy young soldiers, athletes and workers in high heat stress occupations may be victims during the prime of life. Rectal temperatures exceed 40°C with resultant widespread functional and tissue damage. Common denominators of these cases are a high metabolic heat load, environmental conditions contributing heat or failing to accept heat dissipation, and failure of thermoregulatory mechanisms which maintain body heat loads within life-sustaining levels.

The elderly population, especially those already beset with chronic debilitating disease, are particularly afflicted. A high mortality is imposed on this group of population. Their inability to defend against heat illness because of prior compromise of the cardiovascular system is understood (Levine, 1969). During heat waves significant additional deaths due to heat-aggravated or heat-precipitated illness, occur among the elderly in their normal living conditions.

In the conditions prevailing during the Makkah Pilgrimage (Hajj), especially during the hot cycle, the elderly are at a very high risk. This paper deals with conditions prevailing during the Hajj and the confounding factors at work which precipitate hundreds of cases of heat stroke. The possibility of preventing the occurrence of heat stroke among the two million pilgrims, the majority of whom are elderly, is discussed.

THE HAJJ

The Hajj is one of the five pillars of Islam. In 1982 more than two million Muslims, ie., nearly two out of every thousand Muslims, offered the pilgrimage. The Hajj is performed to a sanctuary, the Kaaba, which was built about 25 centuries before the birth of the Prophet Mohamed. The Kaaba is the oldest temple of monotheism (Masjid) known on the earth (Abdalla, 1982).

Pilgrims on arrival in Jeddah from all over the world begin the rituals (manasik) of Hajj (Fig. 1). They dress in Ihram dress, which is essential for all pilgrims. For males, it consists of two plain, white, unsewn cloth garments, one covering the lower body, from the waist to below the knees, and the other the upper half to the neck. Females dress in

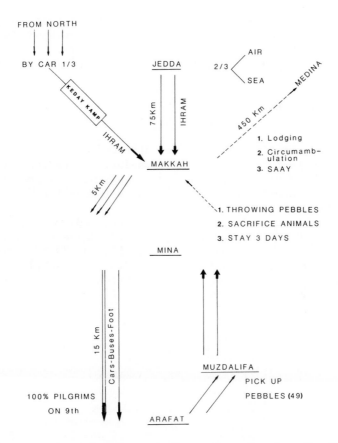

Fig. 1. *Diagrammatic procedure of performing the Hajj rituals.*

white garments, usually with synthetic underwear. On arrival in Makkah, circling or tawaf of the Kaaba is performed. On the eighth day of Dhu Al-Hijja, the pilgrims go to Mina. The next day, the day of Hajj, they travel to Arafat, leaving it after sunset for Muzdalifah (Abdalla, 1982). They return to Mina the next day for ramyee, the stoning of the devils (represented by three stone pillars). The final act of the Hajj is the tawaf al wida, the circling of the Kaaba, before departure. Nearly all pilgrims visit the Prophet's tomb at Medina, 450 km north of Makkah, either before or after the Hajj.

During the Hajj hundreds of thousands of animals - camels, cows, sheep and goats - are slaughtered in a small area within

two days. The Hajj Research Centre has found that about one
million "sheep units" (a cow or a camel is the equivalent of
seven sheep units, and a sheep or goat is one) were sacrificed
in 1982, 70% on the first day of sacrifice half before midday.

THE PILGRIMS

For a long time the Hajj was attended mainly by old people
who wished to fulfil this religious duty before their death.
Still the elderly are predominant among the pilgrims, many of
whom have concomitant chronic diseases (Khogali & Weiner,
1980; Khogali, 1983).

More than two million pilgrims performed the Hajj in 1982;
853,555 from outside and 1,158,000 from inside the Kingdom of
Saudi Arabia. Of the latter group, 919,015 (79.4%) were non-
Saudis, and 238,985 (20.6%) were Saudis. The foreign pilgrims
represented more than 80 different nationalities although 70%
of them came from 10 countries (Table 1).

*TABLE I. Distribution of foreign pilgrims by countries -
Hajj 1402 AH (1982).*

Rank	Country	Number	%
1	Egypt	98,408	11.5
2	Iran	89,503	10.5
3	Nigeria	81,128	9.5
4	Pakistan	72,844	8.5
5	North Yemen	63,241	7.4
6	Indonesia	57,478	6.7
7	Turkey	43,788	5.1
8	Algeria	40,400	4.7
9	Syria	27,890	3.3
10	Sudan	26,983	3.2
Sub Total	Top 10 countries	601,663	70.5
All others	70+ countries	251,892	29.5
Total	80+ countries	853,555	100.0

*Source: Hajj Research Centre - King Abdelaziz University,
Jeddah, Saudi Arabia (1982).*

MODE OF TRAVEL

Seventy three per cent of the foreign pilgrims arrived by air while 20.4% came over land and only 6.5% travelled by sea (Al-Marzoogi *et al.*, 1983).

PREDISPOSING FACTORS

The common denominators predisposing to heat stroke are high environmental temperature and humidity (Stonehill, 1973). Under such conditions the body's ability to lose heat is seriously impaired. Other predisposing factors are strenuous physical exertion, poor physical condition and unacclimatization.

Sanitation, overcrowding, noise and lack of sleep are major problems during the Hajj (Khogali, 1983). Added to this is the fact that performance of the Hajj rituals entails stenuous physical efforts which could be detrimental if practised at high environmental temperature. Direct exposure to the radiant heat of sunlight is an important factor because pilgrims carry out certain rituals with the head uncovered and/or at mid-day, e.g. throwing the pebbles.

In fact the strain on the pilgrims starts at their own home countries, weeks or months before the Hajj. The majority of pilgrims from developing countries are from the rural areas. They have to leave their villages to come to the capital to prepare their travel documents for what might be their first journey by air. By the time they reach the air-port at Jeddah, they are completely exhausted, especially those coming from far away. Their physical fitness and limits of endurance are minimal.

PREVENTION GUIDELINES

Prevention should start in countries of origin, but there must be close co-ordination between different agencies and those in Saudi Arabia. Recommendations for heat stroke and heat illness prevention have changed over past decades and more refinements may evolve.

Success in prevention of heat stroke could be achieved through (i) awareness and education, (ii) acclimatization, (iii) matching the level of activity to ambient temperature and humidity, (iv) liberal water replacement, (v) use of proper clothing, (vi) appropriate personal history and physical examination (Johnson, 1982).

Awareness and Education

Each pilgrim and mutawif (group guide), as well as medical personnel and religious leaders, should be made aware of the risks of heat stroke, and the measures that can prevent it. They should also be made aware of the work load and effort the performance of the Hajj rituals entails and be educated in how to carry them at minimum workload at the most appropriate time.

Acclimatization

Acclimatization is an adaptive process which results in a diminution of physiologic strain produced by repeated application of an environmental stress. With heat acclimatization, sweating is initiated at a lower core temperature and the sweat rate is increased. Sodium conservation is increased due to aldosterone secretion. There is an increase in metabolic efficiency and thus there is better energy utilization. Another aspect of acclimatization is haemodynamic, which augments the ability to increase blood flow both to muscles where heat is generated and to skin where it is dissipated (Martin *et al.*, 1974).

These changes begin to occur early in the course of physical conditioning but may take as long as two months to become complete. As a minimum guideline, acclimatization should start with a short period of conditioning every day and be gradually increased for at least 4-6 weeks.

Level of Activity and Ambient Temperature

In hot climates, physical exertion during the hottest hours should be avoided or at least minimized. A large number of pilgrims perform the Hajj rituals during the hottest hours of the day, especially in Mina. Physical exertion is also associated with going to and from the Kaaba and saying the prayers at mid-day.

Water Replacement

A fact which should also be made common knowledge is that
man, when subjected to heat stress, will not drink sufficient
to replace evaporative water loss, but will suffer from
'voluntary' dehydration (Shibolet *et al.*, 1976), leading to
significant dehydration and predisposing to heat stroke.
Over-hydration has been found to be beneficial during
physical exercise in the heat (Moroff & Bass, 1965).

Free access to water is mandatory. Ideally, cold water
should be provided since it is absorbed more quickly than
warm water (Johnson, 1982). The primary mechanism of heat
stroke involves fluid loss and it is essential, therefore,
that fluid loss be replaced. The use of water and salt
tablets is not recommended - because there seems to be no
justification for it in the acclimatized subject. In
addition, extra salt intake without adequate water intake, can
result in dangerous complications.

Clothing

Up to 70% of the cooling effect of evaporation can be lost
when clothing inhibits air convection, radiation and evaporation (see Robertshaw, 1983). Male pilgrims dress in very
hygienic cotton clothes which seem to facilitate evaporation.
The only shortcoming is that their heads are not covered at
all times. The clothing of women is 'unphysiological',
particularly synthetic underwear which interferes with
evaporation.

Personal History and Physical Examination

An appropriate history and physical examination is of
paramount importance. Many conditions predispose to heat
stroke and certain medicines may affect sweating mechanisms.
For all pilgrims, a proper history and physical examination
should be documented at their countries of origin prior to
their departure. Each pilgrim should be equipped with a
certificate together with his passport specifying previous
heat illness, existing illness and the current therapy undertaken. This will facilitate many problems faced by the
health staff in Saudi Arabia when presented with heat stroke
cases. It will help in the proper management, recovery and
follow up of patients.

DISCUSSION

During Hajj 1982, 1119 cases of heat stroke were admitted
to heat stroke treatment centres along the pilgrimage route.
Their characteristics, age, nationalities and clinical
presentation have been discussed (Khogali *et al.*, 1983;
Alkhawashki *et al.*, 1983). On a very conservative estimate
another 800 cases of heat stroke died before reaching the
heat stroke treatment centres. Essentially all the 2000
cases of heat stroke occurred among the 853,555 foreign
pilgrims, yielding a very high incidence of 2.3 per 1000.
For each case of heat stroke there are 5-8 cases of heat
exhaustion (Ellis, 1976).

Heat stroke can occur under milder climatic conditions
than prevailing in the hot summer in Saudi Arabia. The risk
of heat stroke increases as a hyperbolic function of the Wet
Bulb Globe Temperature (WBGT) index (Wyndham, 1965; Bartley,
1977). Many factors were reported to increase risk of heat
stroke: unacclimatization, dehydration, previous illness,
lack of sleep, poor physical conditioning, overweight and
differences in biological susceptibility. It is most unlikely
that the pilgrims will have been screened for the various risk
factors. Moreover, some risk factors cannot be completely
screened for ahead of time and some develop or worsen with
time, particularly during performance of the Hajj rituals
(eg., dehydration). On the other hand, the metabolic heat
loads from physical exertion and other factors may be great
(Bartley, 1977). With 2 million people crowded in the same
area, the amount of heat generated from their bodies is
tremendous; add to this, the water spilled, and local climatic conditions are drastically aggravated.

It is our belief that the cumulative effect of all the
variables operating during the Hajj, as discussed here and
elsewhere (El-Halawani, 1964; Khogali, 1983; Khogali &
Alkhawashki, 1981), make it extremely unlikely, if not
impossible, that all heat stroke cases can be prevented.
Even in the military world, total prevention of heat stroke
cases has been very difficult to achieve (Minard *et al.*, 1957;
Bartley, 1977). Minard stated that one of the unsolved
problems is the continued occurrence of heat stroke, the
relative rate of which appears to have increased as the
incidence rate of milder forms of heat illness has declined.

Since all heat stroke cases are not preventable under the
operational conditions of the Hajj, it is essential to have
adequate treatment systems available. The use of the word
system for screening heat illnesses was devised by Bartley

(1977) because it takes more than just treatment within a hospital to ensure that all lives possible are saved and morbidity reduced. Such a system has been developed and its main features are outlined elsewhere (Al-Marzoogi *et al*., 1983). This systematized programme was developed in 1981 and it included: emphasis on the seriousness of heat stroke; its recognition and first aid treatment provided at field posts and during ambulance transport; a standard regimen of treatment and operating procedure guide in the heat stroke treatment centres and in the health centres (Khogali, 1982).

In conclusion, it is our belief that the prevention guidelines outlined above, if adhered to, would greatly reduce the number of heat stroke cases and heat-induced illnesses. However, with the many confounding factors characteristic of the Hajj, there will always be many cases of heat stroke; for these cases the second line of attack, viz, an adequate treatment system, is adopted.

Acknowledgement

The authors wish to thank the Minister of Health, Saudi Arabia and all the medical staff of the Ministry of Health who are the corner stone in the first and second lines of prevention. This work is supported by Kuwait University Research Council Grant No. MC 009.

REFERENCES

Abdalla, A. (1982). The Hajj. Arabia: *The Islamic World Review 1*, 22-32.
Alkhawashki, M., Mustafa, M.K.Y., Khogali, M. & ElSayed, H. (1983). Clinical presentation of heat stroke cases during the Makkah pilgrimage. *In* "Heat Stroke and Temperature Regulation" (M. Khogali & J.R.S. Hales, eds), pp. 99-108. Academic Press, Sydney.
Al-Marzoogi, A., Khogali, M. & El-Ergesus, A. (1983). Organizational set up for detecting, screening, treatment and follow-up of heat disorders. *In* "Heat Stroke and Temperature Regulation" (M. Khogali & J.R.S. Hales, eds), pp. 31-39. Academic Press, Sydney.
Bartley, J.D. (1977). Heat stroke: Is total prevention possible? *Milit. Med. 142*, 528-535.
El-Halawani, A.W. (1964). Heat illness during the Mecca Pilgrimage. *WHO Chron. 18*, 283-288.

Ellis, F.P. (1976). Heat Illness. *Trans. Roy. Soc. Trop. Med. Hyg. 70*, 402–425.

Johnson, L.W. (1982). Preventing heat stroke. *Am. Family Pract. 26*, 137–140.

Khogali, M. (1982). Heat disorders with special reference to Makkah Pilgrimage. Monograph. Ministry of Health, Saudi Arabia.

Khogali, M. (1983). Epidemiology of heat illnesses during the Makkah Pilgrimage in Saudi Arabia. *Int. J. Epid.* (in press).

Khogali, M. & Al-Khawashki, M. (1981). Heat stroke during the Makkah Pilgrimage (Hajj). *Saudi Med. J. 2*, 85–93.

Khogali, M. & Weiner, J.S. (1980). Heat stroke: Report on 18 cases. *Lancet ii*, 276–278.

Khogali, M., ElSayed, H., Amar, M., El Sayad, S., Al Habashi, S. & Mutwali, A. (1983). Management and therapy regimen during cooling and in the recovery room at the different heat stroke treatment centres. *In* "Heat Stroke and Temperature Regulation" (M. Khogali & J.R.S. Hales, eds), pp.149–156. Academic Press, Sydney.

Knochel, J.P. (1974). Environmental heat illness. *Arch. Int. Med. 133*, 841–864.

Levine, J.A. (1969). Heat stroke in the aged. *Am. J. Med. 47*, 251–258.

Martin, D.W., Watts, H.D. & Smith, L.H. (1974). Heat stroke: Medical staff conference. *Western J. Med. 121*, 305–312.

Minard, D., Belding, H.S. & Kingston, J.R. (1957). Prevention of heat casualties. *J. Am. Med. Assoc. 165*, 1813–1818.

Moroff, S.V. & Bass, D.E. (1965). Effects of over hydration on man's physiological responses to work in the heat. *J. Appl. Physiol. 20*, 267–270.

Robertshaw, D. (1983). Contributing factors to heat stroke. *In* "Heat Stroke and Temperature Regulation" (M. Khogali & J.R.S. Hales, eds), pp. 13–29. Academic Press, Sydney.

Shibolet, S., Lancaster, M.C. & Danar, Y. (1976). Heat stroke: A review. *Aviat. Space & Environ. Med. 47*, 280–301.

Stonehill, R.B. (1973). Heat stroke. *J. Indiana State Med. Assoc. 66*, 377–379.

Wyndham, E.H. (1965). A survey of the causal factors in heat stroke and of their prevention in the Gold Mining Industry. *J.S. African Inst. Min. Met. 66*, 125–155.

Index

3 4 5 6 7 8 9 0 1 2
A B C D E F G H I J